WITHDRAWN

DESIGNING EFFICIENT ALGORITHMS FOR PARALLEL COMPUTERS

McGraw-Hill Computer Science Series

DESIGNING EFFICIENT ALGORITHMS FOR PARALLEL COMPUTERS

Michael J. Quinn
University of New Hampshire

McGraw-Hill Book Company

New York   St. Louis   San Francisco   Auckland   Bogotá   Hamburg   London
Madrid   Mexico   Milan   Montreal   New Delhi   Panama   Paris   São Paulo
Singapore   Sydney   Tokyo   Toronto

This book was set in Times Roman by Aldine Press.
The editors were Kaye Pace and Larry Goldberg;
the cover was designed by Albert Cetta;
the cover photograph was taken by A. Diakopoulos/Stock, Boston;
the production supervisor was Diane Renda. The drawings were done by ANCO/Boston.
R. R. Donnelley & Sons Company was printer and binder.

Figure Credits

Figures 2-1 and 4-18: Aho, Hopcroft, and Ullman, *The Design and Analysis of Computer Algorithms,* ©1974, Addison-Wesley, Reading, Massachusetts, page 5, Figure 1.3, and page 264, Figure 7.5. Reprinted with permission.

Figures 2-6 and 2-8: Reprinted from *Computational Aspects of VLSI* by Jeffrey D. Ullman, ©1984, with the permission of the publisher, Computer Science Press, Inc., 1803 Research Blvd., Rockville, MD 20850 USA.

Figures 2-11, 2-12, 2-13, and 2-14: Reprinted from *Parallel Computers* by R. W. Hockney and C. R. Jesshope, ©1981, with the permission of the publisher, Adam Hilger, Bristol, England.

Figure 4-9: Robert Sedgewick, *Algorithms,* ©1983, Addison-Wesley, Reading, Massachusetts, page 465 (figure). Reprinted with permission.

Figure 4-12: Donald E. Knuth, *The Art of Computer Programming, Volume 3, Sorting and Searching,* ©1973, Addison-Wesley, Reading, Massachusetts, page 237, Figure 56. Reprinted with permission.

DESIGNING EFFICIENT ALGORITHMS FOR PARALLEL COMPUTERS

2 3 4 5 6 7 8 9 0   DOCDOC   8 9 4 3 2 1 0 9 8 7

ISBN 0-07-051071-7

Library of Congress Cataloging-in-Publication Data

Quinn, Michael J. (Michael Jay)
  Designing efficient algorithms for parallel computers.

  (Supercomputing & artificial intelligence)
  Bibliography; p.
  Includes index.
  1. Parallel processing (Electronic computers)
2. Algorithms. I. Title. II. Series.
QA76.5.Q56 1987   004'.35   86-27230
ISBN 0-07-051071-7

To VICTORIA,
my wife and best friend

CONTENTS

PREFACE

A few years ago parallel computers could be found only in research laboratories. Now they are available commercially. For this reason we are entering an exciting period, when work on parallel algorithms can progress beyond design and analysis into implementation and use. This book has two primary goals: to familiarize the reader with classical results and to provide practical insights into how algorithms are made to run efficiently on processor arrays, multiprocessors, and multicomputers.

Chapter 1 puts parallel computing in perspective, showing the need for higher-performance computers and summarizing methods used in the past to increase computer performance. Chapter 2 begins with a presentation of a number of fundamental processor organizations and continues with a description of three parallel computer architectures: the processor array, the multiprocessor, and the multicomputer. Chapter 3 addresses many of the efficiency issues confronting the designer of parallel algorithms.

The chapters after Chapter 3 are more specialized. Chapter 4 presents some important results in parallel sorting from the large body of work done in this area. Chapter 5 discusses dictionary operations and illuminates trade-offs between the complexity of the underlying sequential algorithm and the potential for keeping a large number of processors busy doing useful work. Matrix multiplication is a fundamental component of many numerical and nonnumerical algorithms. Results in parallel matrix multiplication appear in Chapter 6. Chapter 7 describes parallel numerical algorithms to solve recurrence relations, partial differential equations, and systems of linear equations. Chapter 8 surveys parallel algorithms for searching graphs and finding connected components, minimum spanning trees, and shortest paths in graphs.

The final three chapters address current trends and past successes. Areas in which parallel computing may have a significant impact in the future include artificial intelligence and logic programming. Chapter 9 describes potential parallelism in the solution of combinatorial search problems. These problems occur in artificial intelligence, operations research, and graph theory, among other areas. Chapter 10 introduces Prolog, a logic programming language, and summarizes approaches to executing logic programs in parallel. Pipelined vector processors have been of great historical importance; two well-known machines, the Cray-1 and Cyber-205, are surveyed in Chapter 11.

The principal audience for this text is intended to be seniors and graduate students in computer science. Suggested prerequisites are calculus, high-level language programming, data structures, operating systems, computer architecture, and the analysis of algorithms.

The book includes many parallel algorithms written in a machine-independent, high-level pseudocode. Experimental results from implementations of parallel algorithms have been included wherever possible. Important results have been presented as theorems, to make them easier to reference. Each chapter ends with a set of exercises. They range from the elementary to the difficult. A Glossary of parallel computing appears after Chapter 11. References are given throughout the text, and a large bibliography appears at the end of the book. A solutions manual is available to instructors only.

I have taught a one-semester graduate-level course in parallel computing at the University of New Hampshire, using earlier drafts of this book. I recommend that you supplement the exercises with actual programming assignments on a parallel computer or a simulator. Programming a parallel computer is a new, difficult, and exciting experience for most students, and they learn a great deal from their efforts. In addition, graduate students should read recent journal articles and conference papers. With these supplements, there is more than enough material for a one-semester course, giving the instructor some latitude. I have usually taken an "historical" approach, covering Chapters 1, 11, 2, and 3 before the midterm examination and Chapters 4, 5, 6, 8, 9, and 10 in less depth after the midterm.

Kai Hwang, B. Jayaraman, and Vipin Kumar provided many helpful suggestions that led to a substantial improvement in the quality of the text between the first and second drafts. Kaye Pace, my editor at McGraw-Hill, always made me feel as if I were her only responsibility. Let me extend my thanks to everyone involved in the production of the book.

I feel fortunate to have had as my dissertation advisor Narsingh Deo, who introduced me to the area of parallel algorithms. I am grateful to Donald Knuth and numerous unknown support people, who made the TEX typesetting system public, and to L. Michael Gray, who installed and maintained the TEX environment at the University of New Hampshire. I had felt for some time that there was a book in me, waiting to get out. Seeing my words transformed into beautifully typeset output was all the catalyst I needed.

Finally, I would like to thank my teachers throughout the years who provided me with such an inspiring example.

Michael J. Quinn

INTRODUCTION

This book is concerned with **parallel computing**, the process of solving problems on parallel computers. Parallel computing is a relatively young field: the Illiac IV, a processor array, became operational in 1975; the first Cray-1, a pipelined vector processor, was delivered in 1976; and low-cost multiprocessors were not available before 1984. The advent of very large-scale integration (VLSI) heralded a new era in computing: not only did it make the personal computer possible, but also it made practical the development of large-scale computing devices consisting of tens, hundreds, even thousands of processors, all working together to perform a computation.

Although the study of parallel computing is a new discipline, it is far from unimportant. Many programs that run well on conventional computers are not easily transformed to programs that efficiently harness the capabilities of parallel computers. Conversely, algorithms that are less efficient in a sequential context often reveal an inherent parallelism that makes them attractive bases for parallel programs.

Many claim we are entering "the decade of the parallel computer." Applications demand computers that are many *orders of magnitude* faster than the fastest computers available today. Parallelism represents the most feasible avenue to achieve this kind of breakthrough, and countries throughout the western world are vigorously developing ever more powerful parallel computers. Some of these computers are quite expensive: the Connection Machine, marketed by Thinking Machines Corporation, contains up to 65,536 processors and costs $3 million. Other parallel computers cost little more than professional work stations. Low-cost multiprocessors and multicomputers have been announced by Ametek, Aretè, Encore, Intel, NCUBE, Sequent, and other companies.

1

The remainder of this chapter puts the problem of parallel computing in context. Section 1-1 lists a few applications that demand computers much faster than those presently available. Section 1-2 presents a brief history of architectural advances used to increase the performance of computers over the past 35 years. This section also explores the difference between pipelining and parallelism. Section 1-3 introduces the architectural classification schemes of Flynn and Händler. Finally, Section 1-4 examines reasons that have traditionally been given opposing the feasibility of high-level parallel computation. Some of these reasons can now be refuted easily; others are more weighty. It is good to remember that it takes a special kind of creativity to unleash the full power of a parallel architecture. Perhaps that is the best reason to study parallel computing: Success is more difficult and hence more rewarding.

1-1 THE NEED FOR HIGHER-PERFORMANCE COMPUTERS

The increasing power of computers has led to greater visions of what they might be able to do. Hwang and Briggs [1984] point out that mainstream computer usage is gradually becoming more and more sophisticated, progressing from data processing and information processing to knowledge processing and, eventually, intelligence processing. Each level of increasing sophistication demands much more powerful computers. We consider a few examples of current applications that could use extremely powerful computers. Although most of these applications are using computers for "number crunching," other applications are using computers to manipulate symbols or ideas. Hwang and Briggs [1984] are the primary source of information for these examples; their text contains more information about these and other uses.

Weather Prediction

Forecasting the weather on a computer requires the solution of general circulation model equations in a spherical coordinate system. A three-dimensional grid partitions the atmosphere by altitude, latitude, and longitude. Time is the fourth dimension; it, too, is partitioned by specifying a time increment. Given a grid with 270 miles on a side and an appropriate time increment, about 100 billion operations must be performed to compute a 24-hour forecast. This can be done in about 100 minutes on a computer capable of performing 100 million operations per second, such as a Cray-1. A grid this coarse is capable of producing a forecast for New York and Washington, D.C., but not for Philadelphia, located approximately halfway between the other two cities. To get a more accurate forecast for Philadelphia, the grid size would have to be halved in all four dimensions, leading to a 16-fold increase in the number of computations required. A computer capable of 100 million floating-point operations per second (100 megaflops), like the Cray-1, would require 24 hours to complete the 24-hour forecast. Even this new grid size would not be sufficiently fine to allow reliable long-range forecasting. If we want to receive accurate long-range forecasts, much more powerful computers must be developed.

Computational Aerodynamics

Wind tunnel experiments have a number of fundamental limitations. These include the model size, wind velocity, density, temperature, wall interference, and other factors. Numerical flow simulations have none of these limitations. The replacement of wind tunnels by computers has been limited only by the processing speed and memory capacity of the computer being used. The Burroughs Corporation and Control Data Corporation have proposed supercomputers, known as the numerical aerodynamic simulation facilities, with the goal of eliminating the need for wind tunnels. These supercomputers are designed to perform more than a billion floating-point operations per second (gigaflops).

Researchers at the University of Illinois, supported by a grant from the National Science Foundation, have begun using supercomputers to study wind shear. Using computational aerodynamics to "fly" simulated airplanes through microbursts, they hope to learn more about microbursts and the dangers posed to commercial aviation [USA Today 1985].

Artificial Intelligence

Most current computers have a relatively inflexible input/output (I/O) interface. If computers are to become more "user-friendly," they must be able to interact with humans at a higher level, using speech, pictures, and natural language. Allowing voice, pictorial, and natural language input to be handled in real time requires an enormous amount of computing power, much more than is available on standard architectures.

Japan has begun a project to develop fifth-generation computers. One of the goals of the project is to build a computer capable of making 100 million to 1 billion logical inferences per second. Since one logical inference may take anywhere from 100 to 1000 machine instructions to execute, such a machine would have to be able to perform between 10 billion and 1 *trillion* instructions per second.

Remote Sensing

The analysis of earth-resource data broadcast from satellites has many applications in agriculture, ecology, forestry, geology, and land use planning. However, images are often so large that even simple calculations require large amounts of CPU time. For example, a single Thematic Mapper image from the latest Landsat satellite is a 6000-picture element (pixel) by 6000-pixel square. Each pixel is represented by 8 bits; the entire picture is made up of eight images, or bands. A single picture, then, is represented by 288 megabytes of information.

NASA has installed the Massively Parallel Processor (MPP), manufactured by Goodyear Aerospace, to perform satellite image processing.

Nuclear Reactor Safety

The importance of being able to do simulations in real time is evident when you consider the problem of providing computer-aided analysis and simulation of events in a nuclear reactor. The ability to double-check a corrective measure before it is taken could help keep minor malfunctions from turning into major catastrophes. Only supercomputers have the processing speed that would enable such calculations to be done in real time.

Military Uses

Many existing supercomputers are being used by agencies doing research for the military. These agencies use supercomputers to design nuclear weapons, simulate their effects, gather intelligence, and process cartographic data in order to generate maps automatically. Since the U.S. Department of Defense continues to support research into parallel computing, clearly it desires even faster computers.

1-2 METHODS USED TO ACHIEVE HIGHER PERFORMANCE

There has always been a demand for faster computers, and computer engineers have used two methods to achieve higher performance. First, they have increased the speed of the circuitry; second, they have increased the number of operations that can take place concurrently, through either pipelining or parallelism. This section highlights the advances made since the inception of the electronic computer, showing how pipelining and parallelism have worked their way into the design of modern high-performance computers. The book by Hockney and Jesshope [1981] serves as the primary source of information on these advances.

Figure 1-1 illustrates how computer performance has increased over the past three decades. There has been roughly a 10-fold increase in the speed of computer arithmetic every 5 years. For example, the speed of floating-point multiplication increased by a factor of about 1,000,000 between the early 1950s and the early 1980s. Only some of the increase can be attributed to an increase in the speed of components. Architectural advances must be credited with the remainder of the increase. For example, the clock of the MPP is only about 1 order of magnitude faster than the clock of the EDSAC1; the rest of the speedup is due to the ability of the MPP to perform 16,384 multiplications in parallel.

The speed of light puts a ceiling on the speed at which electronic components of a certain size can operate. Hence parallelism has been introduced at many levels to improve the performance of computers. Many different architectural

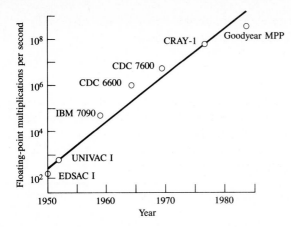

Figure 1-1 The performance of computers has increased by a factor of 10 every 5 years.

advances have been made over the past three decades. Some of these advances are summarized below. Frequently an advance was only possible because a new technology became available. Such new technologies will be mentioned where appropriate.

Bit-Parallel Memory and Bit-Parallel Arithmetic

The first electronic digital computers used a bit-serial main memory. Each bit of a word was read individually from memory. The EDSAC, SEAC, Pilot ACE, EDVAC, and UNIVAC all had mercury delay line (ultrasonic) bit-serial memories.

The first memory to allow all the bits in a word to be accessed in parallel was a cathode ray tube (CRT) system named after F. C. Williams of Manchester University in England. Although Williams used the CRT in bit-serial mode, the SWAC computer at the Institute for Numerical Analysis in England used Williams tubes in a parallel mode with the kth bit of memory words stored in the kth CRT. Williams tubes in parallel mode were also used in the computer at the Institute for Advanced Study and in the IBM 701 (1953).

Bit-parallel arithmetic became possible once bit-parallel memory was available. The IBM 701 was the first commercial machine to have bit-parallel arithmetic.

A prototype ferrite core memory was constructed by Jay Forrester at M.I.T. in 1950. The IBM 704 (1955) was the first commercial computer to use core memory. Besides allowing bit-parallel memory access, cores had the advantages of zero standby power, reasonable cost, speed for general-purpose applications (especially in large systems), and nonvolatility.

I/O Processors (Channels)

In first-generation computers I/O instructions were executed by the CPU. I/O devices continued to become faster (a magnetic tape drive is about 100 times

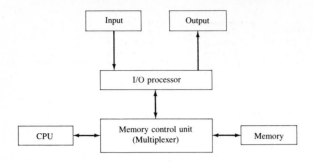

Figure 1-2 I/O processors are one of the architectural advances distinctive of second-generation computers.

faster than a card reader), but even the data transmission speed of a tape drive was far slower than the data manipulation speed of a processor. Because the electromechanical I/O devices were much slower than the electronic CPU, the CPU spent most of its time idling while executing an I/O instruction. This inefficiency frequently was a cause of poor performance [Hayes 1978].

The problem was solved by introducing a separate processor to handle I/O operations. This I/O processor, called a **channel**, receives I/O instructions from the CPU but then works independently, freeing the CPU to resume arithmetic processing. The channel has its own instruction set, custom-tailored for I/O operations. Six channels were added to the IBM 704 in 1958; the new computer was renamed the IBM 709. I/O processors are one of the architectural advances distinctive of second-generation computers (Figure 1-2).

Interleaved Memory

An **interleaved memory** is a memory unit divided into a number of **modules**, or **banks**, that can be accessed simultaneously. Each memory bank has its own addressing circuitry. Instruction and data addresses are interleaved to take advantage of the parallel fetch capability. With **low-order interleaving** the low-order bits of an address determine the memory bank containing the address; with **high-order interleaving** the high-order bits of an address determine the memory bank. Figure 1-3 illustrates the difference between low-order interleaving and high-order interleaving.

When a computer is being designed, it is important to match the speeds of the various components. For example, it does no good to have an extremely fast CPU if the memory unit cannot keep it supplied with instructions and data. The IBM STRETCH computer (1961) was the first computer to have an interleaved memory. Memory interleaving enabled a relatively slow magnetic-core memory to keep up with a fast processor. Memory was divided into two memory banks. Thus, the maximum data transfer rate to and from memory was increased by a factor of 2.

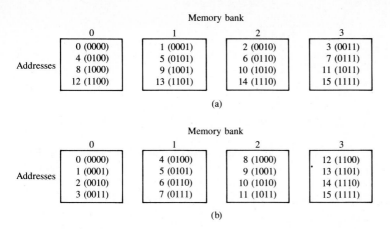

Figure 1-3 Memory interleaving. (a) Low-order interleaving lets the low-order bits of an address determine the memory bank. (b) High-order interleaving lets the high-order bits of an address determine the memory bank.

The ATLAS computer (1963), famous for its innovations in virtual memory, paging, and multiprogramming, had a four-way interleaved memory. The CDC 6600 (1964) divided memory into 32 independent banks. Since the IBM STRETCH, virtually all large computers have used interleaved memory.

Cache Memory

A **cache memory** is a small, fast memory unit used as a buffer between a processor and primary memory. The purpose of a cache memory is to reduce the time the processor must spend waiting for data to arrive from the slower primary memory. The efficiency of a cache memory depends, in part, on the **locality of reference** in the program being run. **Temporal locality** refers to the observed phenomenon that once a particular data or instruction location is referenced, it is often referenced again in the near future [Madnick and Donovan 1974]. **Spatial locality** refers to the observation that once a particular memory location is referenced, a nearby memory location is often referenced in the near future [Madnick and Donovan 1974]. Given a reasonable amount of locality of reference, the majority of the time the processor can fetch instructions and operands from cache memory, rather than primary memory. Only when the instruction or operand is not in the cache memory must the processor idle. Although cache memories began to appear in computers in the early 1960s, they were not economical before the introduction of large-scale integration (LSI) semiconductor memories in the late 1960s. A typical memory hierarchy, showing the position of cache memory, appears in Figure 1-4.

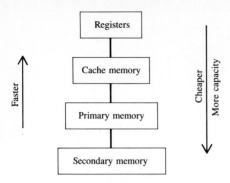

Figure 1-4 A typical memory hierarchy. Registers are the fastest; they are also the most expensive, hence have the smallest capacity.

Instruction Look-Ahead

The use of a cache memory is one solution to the problem of keeping a fast instruction unit busy. Another solution is **instruction look-ahead**, also called **instruction buffering**. In this approach the next few instructions to be executed are always available to the instruction unit, so that instruction fetch can be done as quickly as other operations, such as instruction decode.

The IBM STRETCH computer picked up instructions, decoded them, calculated addresses of operands, and fetched operands several instructions in advance. The IBM 360/91 had the same look-ahead facility. In addition, it was able to prefetch a branch target instruction and some subsequent instructions, to keep the pipe from emptying after every conditional branch. Instruction look-ahead is also needed when there are multiple functional units.

Multiple Functional Units

One way to reap the fruits of instruction and operand prefetching is to examine a set of instructions about to be performed and determine what operations can be performed in parallel without altering the semantics of the program. Given a set of functional units (adders, multipliers, etc.) and precedence relations among a set of operations, the goal is to schedule the operations on the functional units in such a way that the total time needed to perform all the operations is minimized.

CDC 6600 (1964). The CDC 6600 incorporated a number of state-of-the-art features, including the use of 10 functional units to do the actual computations. Two functional units performed increments, two performed floating-point multiplications, and one functional unit was provided for each of the following operations: shift, boolean, branch, floating-point addition, integer addition, and floating-point division. A special control unit called the *scoreboard* selected the registers and functional units to be used in the execution of each instruction. Once a functional unit began processing an instruction, it worked independently of the other functional units. Hence a number of functional units could execute different instructions simultaneously. At least two or three functional units

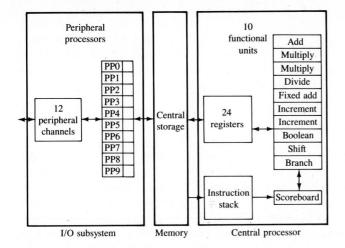

Figure 1-5 Block diagram of the CDC 6600, a third-generation computer. Multiple functional units could be in execution simultaneously. (Reprinted by permission of Control Data Corporation.)

could be expected to be in operation at the same time at any point in the execution of a typical program. Figure 1-5 is a block diagram of the CDC 6600, a third-generation computer.

Instruction Pipelining

Instruction pipelining is the use of pipelining to allow more than one instruction to be in some stage of execution at the same time. Instruction pipelining is achieved by dividing the execution of each instruction into a number of phases, such as fetch, decode, operand fetch, and instruction execute.

Ferranti ATLAS (1963). The ATLAS computer was based on an accumulator architecture; hence instructions were of the one-address type. Execution of instructions was pipelined. The four stages were instruction fetch, operand address calculation, operand fetch, and the arithmetic operation. Pipelining decreased the average time per instruction from 6.0 to 1.6 microseconds. The ATLAS used a 32K word cache as a high-speed buffer between the slow main memory and the arithmetic registers. Without a cache, memory could not keep up with the CPU.

Amdahl 470 V/6 (1975). The Amdahl 470 V/6, introduced in 1975, was the first computer to use LSI technology for logic circuits of the CPU. This made the computer about one-third the size of the IBM 360/168 and more reliable.

Execution of instructions on the V/6 was divided into 12 cycles. Since a new instruction could be fetched every 2 cycles (64 nanoseconds), up to six instructions could be in execution simultaneously. The pipeline was kept full by using a 16-kilobyte cache with an access time of 65 nanoseconds. Bottlenecks to high performance occurred when an instruction took more than 2 cycles in the arithmetic unit or when a cache miss occurred during an instruction or operand fetch, resulting in a block transfer from the slower main memory.

Pipelined Functional Units

CDC 7600 (1969). The CDC 7600 was about 4 times faster than the CDC 6600. The increased speed was mainly due to a reduction in the clock period from 100 to 27.5 nanoseconds. However, the CDC 7600 differed from the CDC 6600 in more than its clock cycle time. The 10 serially organized functional units of the 6600 were replaced by eight pipelined functional units and a serial unit for division (which could not be pipelined). Because the pipelined units were so much faster, it was no longer necessary to duplicate the units for incrementing and multiplying. The additional functional unit was used for counting the number of 1 bits in a word.

Data Pipelining

A **vector computer** is a computer with an instruction set that contains operations on vectors as well as scalars. There are two classical ways of implementing a vector computer. One way is to build a **processor array**, in which each vector element is manipulated by a different arithmetic processor. Processor arrays are described in the next subsection. The second way to implement a vector processor is through pipelining.

Pipelining the data stream is a natural evolution from the traditional serial model of computation. Instead of fetching scalars from memory and performing arithmetic on them, vectors are streamed from memory into the CPU, where pipelined arithmetic units manipulate them. Most supercomputers are **pipelined vector processors**, such as the Cray-1, manufactured by Cray Research, and the Cyber-205, made by Control Data Corporation. These computers are capable of performing hundreds of millions of floating-point operations per second. Pipelined vector processors are surveyed in Chapter 11.

Processor Arrays

A **processor array** is a set of identical synchronized processing elements capable of simultaneously performing the same operation on different data. Processor arrays are a second way to implement vector computers. To elaborate on the difference between a pipelined vector processor and a processor array, a pipelined vector processor achieves concurrency by passing the vector(s) to be manipulated through a pipelined functional unit, while a processor array achieves concurrency by passing each element of the vector(s) to a different arithmetic unit. Processor arrays are discussed in more detail in Chapter 2.

Multiprogramming, Timesharing, and Multiprocessing

The execution of a typical program consists of an alternating sequence of computations and I/O operations [Hayes 1978]. For example, a program executing in a paged virtual memory system may execute a number of instructions and then initiate an I/O operation. When this operation has been completed, instruction execution may continue until perhaps a page fault occurs. Once again, the CPU must suspend execution of the program until the page has been brought into primary memory. Rather than sit idle during I/O operations, the CPU could start (or resume) execution of another program. A **multiprogrammed** operating system allows more than one program to be in some state of execution at the same time.

Timesharing is a kind of multiprogramming that supports the concurrent execution of many user jobs and allows each user to interact with his or her program in real time.

Hayes [1978] defines a multiprocessor to be "an integrated computer system containing two or more CPUs. The qualification 'integrated' implies that the CPUs cooperate in the execution of programs .... Multiprocessing finds its main application as a method of performance enhancement in computers designed to process many programs simultaneously at rates faster than are economically obtainable with a single CPU." In other words, multiprocessors have traditionally been used to increase the throughput of multiprogrammed operating systems.

More recently there has been interest in using multiprocessors for parallel processing—solving individual computationally intensive problems. In this book we use the term **multiprocessor** to refer to a shared-memory multiple-CPU computer designed for parallel processing. A **multicomputer** is a multiple-CPU computer designed for parallel processing but lacking a shared memory. Various kinds of multiprocessors and multicomputers are surveyed in Chapter 2.

Data Flow Computers

Since the late 1960s much research has been done in the area of data flow computing, a radical alternative to the more traditional architectures discussed earlier in this section. All the computers discussed up until now—single-CPU computers, pipelined vector processors, processor arrays, multiprocessors, and multicomputers—are **control-driven**. In a single-CPU computer, pipelined vector processor, or processor array, a centralized control mechanism determines the order in which instructions are to be executed. The address of the instruction currently under execution can be traced to the program counter. Although instructions may manipulate vectors as well as scalars, the processing of instructions is inherently sequential. Multiprocessors and multicomputers allow more than one instruction stream, but their use of program counters in each CPU indicates that they, too, are control-driven.

Data flow computers, however, eliminate the program counter. Control is decentralized. An instruction is ready to execute as soon as its operands are available. Advocates of the data flow approach argue that once a data flow graph for a computation has been constructed, all the parallelism inherent in that computation—both fine-grain and large-grain—has been exposed. Data flow computations are inherently asynchronous and maximally parallel. Other attributes of the data flow approach are high modularity and freedom from side effects.

Critics of data flow computing argue that the results of data flow analysis can be applied without resorting to building a data flow computer. Data flow computers suffer from a number of obstacles that put a damper on their performance. These obstacles include structure storage, inefficient code, a long instruction execution cycle, and lack of instruction pipelining on the critical path. Even proponents of the data flow approach must admit that no data flow supercomputers have yet been built.

Computer Generations

"The division of computer systems into generations is determined by the device technology, system architecture, processing mode, and languages used" [Hwang and Briggs 1984]. Most authors place computers built before 1980 into one of four generations, although some computers defy easy categorization. The primary sources for the categorization that appears below are Baer [1980], Hayes [1978], and Hwang and Briggs [1984].

First-generation computers used electromechanical relays or vacuum tubes to implement logic and memory. All programming was done in machine language. The EDSAC, SEAC, Pilot ACE, EDVAC, and UNIVAC were all first-generation computers.

Second-generation computers are characterized by their use of transistors, printed circuits, and magnetic core memory. Architecturally, second-generation computers are notable for their use of I/O processors, index registers, and hardware to perform floating-point arithmetic. Assembly languages and the first high-level languages, FORTRAN, ALGOL, and COBOL, were developed for second-generation computers. Batch monitors became available. The IBM 709 was a second-generation computer.

During the third generation, small-scale and medium-scale integrated circuits become common, reducing the size and expense of computer hardware. Semiconductor memory began to replace magnetic core memory. Microprogramming became a popular method of simplifying the design of processors and making them more flexible. Multiprogrammed operating systems became available. Timesharing and virtual memory were innovations that appeared toward the end of the third generation. The IBM System/360 series are well-known examples of third-generation computers.

The fourth generation is characterized by the use of large-scale integration and VLSI to construct logic and memory units. Most operating systems are timeshared and use virtual memory. Vectorizing compilers for pipelined vector processors are available. Commercial supercomputers use a great deal of pipelining and parallelism. The Amdahl 470/V6 and Cray-1 are representative of this generation.

Differences between Pipelining and Parallelism

Most high-performance modern computers exhibit a great deal of concurrency. For example, multiprocessing is a method used to achieve concurrency at the job or program level, while instruction pipelining can be viewed as a method of achieving concurrency at the interinstruction level. However, it is not desirable to call every modern computer a parallel computer. The concurrency of many machines is totally invisible to the user. Hence we adopt the following definitions.

Parallel processing is a kind of information processing that emphasizes the concurrent manipulation of data elements belonging to one or more processes solving a *single problem*. Pipelining and parallelism are two methods used to achieve concurrency. Pipelining increases concurrency by dividing a computation into a number of steps, while parallelism is the use of multiple resources to increase concurrency.

A **parallel computer** is a computer designed for the purpose of parallel processing.

A **supercomputer** is a general-purpose computer capable of solving individual problems at extremely high computational speeds, compared with other computers built at the same time. Of course, *extremely high* is a relative term. By this definition there have always existed supercomputers. Today's standards dictate that a supercomputer must be a parallel computer.

The **throughput** of a device is the number of results it produces per unit time. A pipelined computation is divided into a number of steps, called **segments**, or **stages**. Each segment works at full speed on a particular part of the entire computation. The output of one segment is the input of the next segment. If all the segments work at the same speed, the work rate of the pipeline is equal to the sum of the work rates of the segments, once the pipe is full. Analogies hold between a pipeline and an assembly line: the flow of results is simple and fixed, precedence constraints must be honored, and it takes time to fill and drain the pipe. If we assume that each segment of the pipe requires the same amount of time, the multiplicative increase in the throughput is equal to the number of segments in the pipeline. Figure 1-6 illustrates pipelining in the context of an imaginary automobile assembly line.

Speedup is the ratio between the time needed for the most efficient sequential algorithm to perform a computation and the time needed to perform a computation on a machine incorporating pipelining and/or parallelism. (This rather intuitive definition is made more rigorous in Chapter 2.)

An example should help illuminate the difference between pipelining and parallelism as well as provide a practical demonstration of speedup. Assume that it takes 3 units of time to assemble a widget. Furthermore, assume that this assembly consists of three steps—A, B, and C—and each step requires exactly one unit of time. A sequential widget-assembly machine assembles a widget by spending 1 unit of time performing step A, followed by 1 unit of time performing step B, followed by 1 unit of time performing step C. Clearly a sequential widget-assembly machine produces one widget in 3 time units, two widgets in 6

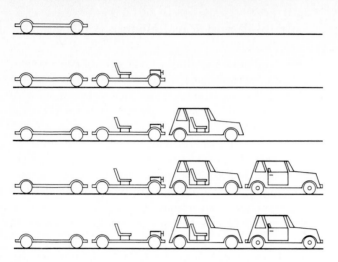

Figure 1-6 An automobile assembly line is an example of a pipeline.

time units, and so on, as shown in Figure 1-7a. Consider how the output could be increased if the assembly were pipelined. Figure 1-7b illustrates a three-segment pipeline. Each of the subassembly tasks has been assigned to a unique machine. The first machine performs subassembly task A on a new widget every time unit and passes the partially assembled widget to the second machine. In a similar way the second machine performs subassembly task B, and the third machine performs subassembly task C. The pipelined widget-assembly machine produces one widget in 3 time units, as does the sequential machine, but after that initial time to fill the pipe (assembly line), one widget appears every time unit. Hence the second widget appears at time 4, the third widget at time 5, and so on. Figure 1-7c shows a group of three parallel widget-assembly machines. Each machine performs every subassembly task, as does the sequential widget assembler. Throughput is increased by replicating machines. Clearly three widgets appear in 3 time units, six widgets appear in 6 time units, and so on. Note that the time needed to produce four widgets is the same as the time needed to produce five or six widgets. (Why?)

Figure 1-8 illustrates the speedup achieved by the pipelined and parallel widget machines. The x axis represents the number of widgets assembled; the y axis represents the speedup achieved. For any particular number of widgets i, the speedup is computed by dividing the time needed for the sequential machine to assemble i widgets by the time needed for the pipelined or parallel machine to assemble i widgets. For example, the sequential machine requires 12 time units to assemble four widgets, while the pipelined machine requires 6 time units to assemble four widgets. Hence the pipelined machine exhibits a speedup of 2 for the task of assembling four widgets.

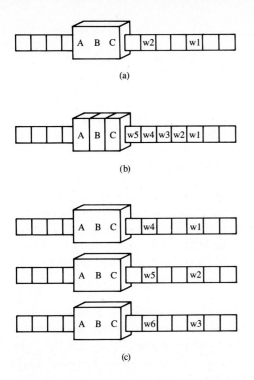

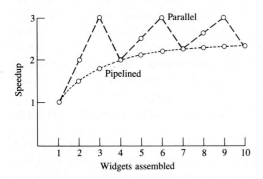

Figure 1-7 Three methods to assemble widgets. (a) A sequential widget-assembly machine produces one widget every 3 units of time. (b) A three-segment pipelined widget-assembly machine produces the first widget in 3 units of time and successive widgets every time unit thereafter. (c) A three-way parallel widget-assembly machine produces three widgets every 3 units of time.

Figure 1-8 Speedup achieved by pipelined and parallel widget-assembly machines. Note that speedup is graphed as a function of problem size (number of widgets assembled).

1-3 CLASSIFYING ARCHITECTURES

This section presents two methods of classifying computer architectures. The first method, published by Flynn [1966], considers the presence or absence of potential multiplicity in the instruction and data streams of a computer. Händler's

[1977] method associates with each computer an expression indicating the pipelining and parallelism to be found at three levels in the hardware of a computer system.

Flynn's Taxonomy

Flynn [1966] bases his taxonomy of computer architectures on the concepts of instruction stream and data stream. An **instruction stream** is a sequence of instructions performed by a computer; a **data stream** is a sequence of data used to execute an instruction stream. Flynn categorizes an architecture by the multiplicity of hardware used to manipulate instruction and data streams. "The multiplicity is taken as the maximum possible number of *simultaneous* operations (instructions) or operands (data) being in the same phase of execution *at the most constrained* component of the organization" (Flynn's emphasis) [Flynn 1966]. Given the possible multiplicity of instruction and data streams, four classes of computers result.

Single-*Instruction* stream, *Single* Data stream (SISD). Most serial computers fall into this category. Although instruction execution may be pipelined, computers in this category can decode only a single instruction in unit time. An SISD computer may have multiple functional units (e.g., CDC 6600), but these are under the direction of a single control unit.

Single-*Instruction* stream, *Multiple* Data stream (SIMD). Processor arrays fall into this category. A processor array executes a single stream of instructions, but contains a number of arithmetic processing units, each capable of fetching and manipulating its own data. Hence in any time unit a single operation is in the same state of execution on multiple processing units, each manipulating different data.

Multiple-*Instruction* stream, *Single* Data stream (MISD). No computers fit into this category.

Multiple-*Instruction* stream, *Multiple* Data stream (MIMD). This category contains most multiprocessor systems. As noted before, early multiprocessors were built to increase the throughput of multiprogrammed operating systems, and there was little interaction among the CPUs. In this book the label *MIMD* is reserved for multiple-CPU computers designed for parallel processing; that is, computers designed to allow efficient interactions among their CPUs.

Flynn's classification scheme is too vague to allow an iron-clad labeling of modern supercomputers. It is only natural, then, that there is some disagreement among experts as to how to classify certain architectures. For example, in what category does the Cray-1, a pipelined vector processor, belong? Because it is a vector computer, Hockney and Jesshope [1981] label the Cray-1 an example of an SIMD architecture. Because it does not have multiple processing elements, Hwang and Briggs [1984] put the Cray-1 into the SISD category. Hwang and Briggs' position seems more consistent with Flynn's definition that was quoted earlier. Thus in this text the terms *SIMD* and *processor array* are synonymous.

Händler's Classification

Händler [1977] has proposed a notation for expressing the pipelining and parallelism occurring at three levels of a computer system: the processor control unit (PCU), the arithmetic logic unit (ALU), and bit-level circuit (BLC). A PCU corresponds to a processor or CPU, an ALU corresponds to a functional unit or a processing element of an array processor, and the BLC corresponds to the logic needed to perform single-bit operations in the ALU.

A computer is described by three pairs of integers:

$$T(C) = \langle K \times K', \ D \times D', \ W \times W' \rangle$$

where K = number of PCUs
K' = number of PCUs that can be pipelined
D = number of ALUs controlled by each PCU
D' = number of ALUs that can be pipelined
W = number of bits in an ALU or processing element (PE) word
W' = number of pipeline segments in all ALUs or in a single PE

If the value of the second element of any pair is 1, it is omitted. The $\times$ operator can be used to link descriptions of different kinds of processors in a single computer system.

To illustrate the notation, consider how the CDC 6600 is classified. The CDC 6600 has a single CPU. The ALU has 10 functional units, and all these may be pipelined together. The word length on a CDC 6600 is 60 bits. Up to 10 peripheral I/O processors may work in parallel with each other and with the CPU. Each I/O processor contains a single ALU with a word length of 12 bits. Hence the CDC 6600 is described as follows:

$$T(\text{CDC } 6600) = T(\text{central processor}) \times T(\text{I/O processors})$$
$$= \langle 1, \ 1 \times 10, \ 60 \rangle \times \langle 10, \ 1, \ 12 \rangle$$

The Cray-1 is a single-CPU computer with 12 pipelined functional units. Different functional units have from 1 to 14 segments. Up to eight functional units can be chained together to form a pipeline. A word on the Cray-1 has 64 bits. Händler's description of the Cray-1 is

$$T(\text{Cray-1}) = \langle 1, \ 12 \times 8, \ 64 \times (1 \sim 14) \rangle$$

1-4 ARGUMENTS AGAINST THE MERITS OF HIGH-LEVEL PARALLELISM

Grosch's law. Grosch's law states that the speed of computers is proportional to the square of their cost [Grosch 1953, 1975]. Thus if you are looking for a fast computer, you are better off spending your money buying one large computer than two smaller computers and connecting them.

Rebuttal of Grosch's law. Grosch's law is true within classes of computers (mainframes, minicomputers, microcomputers), but it is not true between classes. Within any class, such as the class of microcomputers, the power of the computer rises quicker than its cost, but you can buy more megaflops per dollar by spending your money on top-of-the-line microprocessors, rather than pipelined vector processors [Ein-Dor 1985].

Furthermore, computers may be priced according to Grosch's law, but the law cannot be true asymptotically. There is only one fastest computer, and it has a certain price. You cannot get a faster computer by spending more. The only way to get a higher-performance computer is to use a number of computers in parallel.

Minsky's conjecture. Minsky's conjecture states that the speedup achievable by a parallel computer increases as the logarithm of the number of processing elements, thus making large-scale parallelism unproductive [Minsky and Papert 1971].

Rebuttal of Minsky's conjecture. Experimental results prove that the speedup achievable by a parallel computer depends strongly on the particular algorithm and the computer's architecture. Sometimes implementations of parallel algorithms do exhibit logarithmic speedup. Many algorithms, however, have shown linear speedup for over 100 processors.

History. History tells us that the speed of traditional single-CPU computers has increased 10-fold every 5 years. Why should great effort be expended to devise a parallel computer that will perform tasks 10 times faster when, by the time the new architecture is developed and implemented, single-CPU computers will be just as fast?

Rebuttal. First, it seems certain that there will always be a demand for faster computers. If that is the case, then why not invest the resources necessary to develop a good, general-purpose parallel architecture? Once they are built, the speed of these computers can be expected to increase by a factor of 10 every 5 years, too.

Second, many applications demand computers several orders of magnitude faster than existing machines. It will be a long time before sequential computers have enough power to solve these problems. Utilizing parallelism is better than waiting.

Amdahl's law. A small number of sequential operations can effectively limit the speedup of a parallel algorithm. Let f be the fraction of operations in a computation that must be performed sequentially, where $0 \leq f \leq 1$. It is easy to see that the maximum speedup S achievable by a parallel computer with p processors performing the computation is

$$S \leq \frac{1}{f + (1 - f)/p}$$

A corollary follows immediately from Amdahl's law: a small number of sequential operations can significantly limit the speedup achievable by a parallel computer. For example, if 10 percent of the operations must be performed sequentially, then the maximum speedup achievable is 10, no matter how many processors a parallel computer has.

Rebuttal of Amdahl's law. Amdahl's law is one of the stronger arguments against the future of parallel computation, because if parallel computers can never run more than 10, 15, or 20 times faster than single-CPU computers, then the "wait for faster technology" argument makes a lot of sense. However, there do exist some parallel algorithms with almost no sequential operations. Hence Amdahl's argument serves as a way of determining whether an algorithm is a good candidate for parallelization. We explore the implications of Amdahl's law further in Chapter 2.

Pipelined computers are sufficient. Most supercomputers are vector computers, and most of the successes attributed to supercomputers have been accomplished on pipelined vector processors, especially the Cray-1 and Cyber-205. Vectorizing compilers that will find the parallelism in FORTRAN programs and adapt them to a vector processing environment already exist. Converting programs to multiprocessors or multicomputers seems much more difficult.

Rebuttal. If only vector operations can be executed at high speed, supercomputers will not be able to tackle a large number of important problems. The latest supercomputer designs incorporate both data pipelining and high-level parallelism. For example, the ETA[10], the successor to the Cyber-205, and the Cray-2 and Cray-3, successors to the Cray-1, contain a number of pipelined vector processors designed to work in parallel.

Software inertia. Billions of dollars worth of FORTRAN software exists. Who will rewrite it? Virtually no programmers have any experience with a machine other than a single-CPU computer. Who will retrain them?

Rebuttal to argument based on software inertia. This is a backward-looking argument, in the sense that it assumes present and future parallel computers will be solving the same problems that traditional computers solve. A more reasonable assumption is that parallel computing will open up a wide vista of new computer applications. In a few years students will be trained on parallel computers and will be employed trying to solve new computationally intensive problems on these new architectures.

To address the question of existing software, it is reasonable to assume that important software will be rewritten and valuable programmers can be retrained. Work is already underway to develop preprocessors that will take a program written in normal FORTRAN and determine "how much of the program would benefit by being executed in a vector or parallel manner" [ETA 1984].

1-5 SUMMARY

A large number of important applications demand computers several orders of magnitude faster than today's fastest computers. Introduction of high-level parallelism seems to be the surest route to achieving the incredibly high performance demanded by the applications of the 1990s. In the past a number of arguments have been made against parallelism. Some of these arguments can now be refuted easily; others are more serious. Clearly some applications do not have to be run on parallel computers—they run well enough on conventional architectures. A faster solution to some problems may not be possible, even if it is desirable, because the problem does not contain enough inherent parallelism. However, there remain a large body of problems for which solution on a parallel computer seems plausible.

BIBLIOGRAPHIC NOTES

All the references in this book can be found in the Bibliography, which appears after the Glossary and before the Index.

This chapter, by necessity, has surveyed architectural advances quite rapidly. Readers desiring further details can find an excellent study of the evolution of computers in Hayes' book [1978].

Levine [1982], Baer [1980], Bernhard [1982], and Schaefer and Fisher [1982] describe the need for higher-performance computers. A number of books describe parallel computer architectures in more or less detail, including Baer [1980], Hayes [1978], Hwang [1984], Kuck [1978], Lorin [1972], and Stone [1980]. Articles on parallel processing include Harrison and Wilson [1983], Haynes et al. [1982], Kuck [1977], and Kuhn and Padua [1981]. Enslow [1974] and Satyanarayanan [1980] have written books on multiprocessor systems. Hockney and Jesshope [1981] cover processor arrays, attached processors, and pipelined vector processors. An encyclopedic coverage of the major parallel architectures is given by Hwang and Briggs [1984]. Uhr [1984] presents a vision of the future of parallel computing from the point of view of a researcher in artificial intelligence. Frenkel [1986] discusses two modern parallel computers: the Connection Machine of Thinking Machines Corporation and the T Series of Floating Point Systems.

The February 1982 issue of *Computer* is dedicated to data flow computing; it is a good place to start learning more about the subject. Particularly useful articles in this issue are those by Ackerman [1982], Davis and Keller [1982], Gajski et al. [1982], and Watson and Gurd [1981]. Gurd, Kirkham, and Watson [1985] have described the prototype data flow computer built at the University of Manchester.

Much attention has been given to designing parallel algorithms for VLSI circuits. A representative set of references includes Bilardi, Pracchi, and Preparata [1981]; Chazelle and Monier [1981a, 1981b]; Kung [1982]; Mead and Conway [1980]; Lang et al. [1983]; Leighton [1983]; Leiserson [1983]; Shröder [1983]; Thompson [1980, 1983]; and Ullman [1984].

Developments in parallel computer architectures and parallel algorithms continue at an increasing pace. Sources of articles on parallel computing include *Communications of the ACM*, *Computer*, *Future Generations Computer Systems*, *IEEE Transactions on Computers*, *IEEE Transactions on Software Engineering*, *International Journal of Parallel Programming*, *Journal of Parallel and Distributed Computing*, *New Generation Computing*, *Parallel Computing*, *Proceedings of the Annual Symposium on Foundations of Computer Science*, *Proceedings of the Annual Symposium on Computer Architecture*, *Proceedings of the Annual Symposium on Theory of Computing*, and *Proceedings of the International Conference on Parallel Processing*.

EXERCISES

1-1 Try this experiment with a few friends. Shuffle a deck of cards, then determine how long it takes one person to sort the cards into the order A♠, 2♠, ..., K♠, A♡, 2♡, ..., K♡, A♣, 2♣, ..., K♣, A♢, 2♢, ..., K♢. (Is it faster to sort the cards initially by suit or by value?) How long does it take p people to sort p decks of shuffled cards? How long does it take p people to sort 1 deck of cards? Try this experiment for $p = 1, 2, \ldots, 6$.

1-2 Given a task that can be divided into m subtasks, each requiring 1 unit of time, how much time is required for an m-stage pipeline to process n tasks?

1-3 How many widgets must the pipelined widget-assembly machine of Figure 1-7b assemble in order to achieve a speedup of 3 over the sequential machine? Justify your answer.

1-4 How many widgets must the parallel widget-assembly machine of Figure 1-7c assemble in order to achieve a speedup of 3 over the sequential machine? Justify your answer.

1-5 Consider a parallel pipelined widget-assembly machine that has three pipelines, where each pipeline has three segments. Draw the speedup curve for this machine for $1, \ldots, 10$ widgets assembled.

1-6 Define the following: microsecond (μs), millisecond (ms), nanosecond (ns), picosecond (ps), megaword (Mword), megabyte (Mbyte), kiloword (Kword), and kilobyte (Kbyte). A nanosecond is to a second as a second is to how many years?

1-7 What limits are there to the speed of electronic circuits? What is the maximum distance a memory unit could be from an arithmetic unit and still make a 100-picosecond memory access time conceivable?

1-8 What is wrong with the following argument? "The multiplicative increase in the throughput of a pipeline is equal to the number of stages in the pipeline. For example, an adder with five stages, each 1 time unit in length, has 5 times the throughput of a nonpipelined adder taking 5 time units to do an addition. Using the same principle, we could increase throughput even further by creating a 10-stage adder. The first five stages would just pass the data along, while the last five stages would be identical to the five-stage adder described above. Since this adder has 10 stages, its throughput would be 10 times the throughput of a nonpipelined adder."

1-9 Explain why VLSI technology has been both a boon and a bane for advocates of parallel computing.

1-10 Consider the instruction pipelining scheme of the Amdahl 470 V/6. What problems would you encounter while trying to keep the pipe full? How might these problems be solved? What fundamental obstacle keeps the instruction pipeline from remaining full at all times?

1-11 What keeps the amount of concurrency achievable by the Cyber 6600 from being arbitrarily large?

1-12 Decide whether each of the following advances represents an example of pipelining, parallelism, or a technological advance: bit-parallel memory, I/O processors, interleaved memory, cache memory, instruction look-ahead, multiple functional units, instruction pipelining, pipelined functional units, and data pipelining.

1-13 Construct arguments for and against this statement: "Any computer with an I/O channel is a multiprocessor."

1-14 Prove that if $(1/k)$th of the time spent executing an algorithm involves operations that must be performed sequentially, then the maximum speedup achievable by a parallel form of the algorithm is k.

TWO

MODELS OF PARALLEL COMPUTATION

The goal of this chapter is to introduce three important models of parallel computation and some interesting parallel computer designs. The parallel models are processor arrays, multiprocessors, and multicomputers. All of these models have fostered actual parallel computers.

Section 2-1 describes an important model of serial computation, the random access machine, and introduces terminology used to describe the complexity of algorithms. Those familiar with these topics may wish to skip this section. A number of processor organizations are presented in Section 2-2: mesh, pyramid, perfect shuffle, butterfly, hypercube, and cube-connected cycles. Section 2-3 surveys a number of processor array models as well as the ICL DAP, a commercial mesh-connected processor array. Section 2-4 discusses multiprocessors and multicomputers. Three commercial machines are described: the Denelcor HEP; the Bolt, Beranek, and Newman Butterfly; and the Intel iPSC multicomputer. The goal of these sections is to give an idea of what has been done, rather than exhaustively list every implementation.

2-1 A MODEL OF SERIAL COMPUTATION

The **random access machine (RAM)** is a model of a one-address computer. A RAM consists of a memory, a read-only input tape, a write-only output tape, and a program (Figure 2-1). The program is not stored in memory and cannot be modified. The input tape contains a sequence of integers. Every time

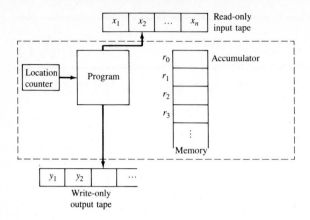

Figure 2-1 The random access machine (RAM) model of serial computation. (Aho, Hopcroft, and Ullman [1974].)

an input value is read, the input head advances one square. Likewise, the output head advances after every write. Memory consists of an unbounded sequence of registers, designated $r_0, r_1, r_2, \ldots$. Each register can hold a single integer. Register r_0 is the accumulator, where the computations are performed.

The exact instructions are not important, as long as they resemble the instructions found on an actual computer. Hence a RAM should have instructions such as **load, store, read, write, add, subtract, test, jump,** and **halt**.

The **worst-case time complexity** of a RAM program is the function $f(n)$ which is the maximum, over all inputs of size n, of the time taken by the program to execute. The **expected time complexity** of a RAM program is the average, over all inputs of size n, of the execution times. Analogous definitions hold for **worst-case space complexity** and **expected space complexity** by substituting the word *space* for *time* in the above definitions.

There are two ways of measuring time and space on the RAM model. The **uniform cost criterion** says each RAM instruction requires 1 time unit to execute and every register requires 1 unit of space. The **logarithmic cost criterion** takes into account the fact that an actual word of memory has a limited storage capacity. The uniform cost criterion is appropriate if the values manipulated by the program always fit into one computer word. That is always the case in this text; hence we always use the uniform cost criterion when computing time and space complexity.

Order Notation. The order notation used in this text is identical to that first proposed by Knuth [1976]. Assume that f and g are functions over the domain of the natural numbers. Then $O(f(n))$ [read "order at most $f(n)$" or "order $f(n)$"] is the set of all $g(n)$ such that there exist positive constants c and n_0 so that $|g(n)| \le cf(n)$ for all $n > n_0$. For example, $4n$ is $O(n^2)$, since for all $n > 1$

it is true that $4n < 2n^2$ (let $c = 2$ and $n_0 = 1$). Similarly $n^2/2$ and $37n^2 + 14n$ are $O(n^2)$.

And $\Omega(f(n))$ [read "order at least $f(n)$"] is the set of all $g(n)$ so that there exist positive constants c and n_0 so that $g(n) \geq cf(n)$ for all $n \geq n_0$. For example, $n/2$ and $4n^2$ are $\Omega(n)$. Likewise, $\Theta(f(n))$ [read "order exactly $f(n)$"] is the set of all $g(n)$ such that $g(n)$ is $O(f(n))$ and $g(n)$ is $\Omega(f(n))$. Hence $n/2$ is $\Theta(n)$, but $4n^2$ is not.

2-2 PROCESSOR ORGANIZATIONS

This section formally defines six important processor organizations—methods of connecting processors in a parallel computer. The primary reference for the material is Ullman [1984], except as noted.

Mesh

In a **mesh network**, the nodes are arranged into a q-dimensional lattice. Communication is allowed only between neighboring nodes; hence interior nodes communicate with $2q$ other processors. Figure 2-2a illustrates a two-dimensional mesh. Some variants of the mesh model allow wrap-around connections between processors on the edge of the mesh. These connections may connect processors in the same row or column (Figure 2-2b), or they may be toroidal (Figure 2-2c).

Efficient sorting and matrix multiplication algorithms have been designed for meshes of processors. We study these algorithms in Chapters 4 and 6. In addition, processor meshes are well suited to solving systems of second-order partial differential equations, as we see in Chapter 7.

Mesh networks do have the disadvantage from a theoretical point of view that data routing requirements often prevent the development of $O(\log^k n)$[†] parallel algorithms. Two examples illustrate this point: sorting n elements on a two-dimensional mesh requires time $\Omega(\sqrt{n})$, and at least $0.35n$ routing steps are needed to compute the product of two $n \times n$ matrices on a two-dimensional mesh [Gentleman 1978].

Pyramid

A **pyramid network of size** p is a complete 4-ary rooted tree of height $\log_4 p$ augmented with additional interprocessor links so that the processors in every tree level form a two-dimensional mesh network [Miller and Stout 1986]. A pyramid of size p has at its base a two-dimensional mesh network containing $p = k^2$ processors. The total number of processors in a pyramid of size p is $\frac{4}{3}p - \frac{1}{3}$. The levels of the pyramid are numbered in ascending order; the base

---

[†]Unless otherwise noted, all logarithms in this book are to base 2.

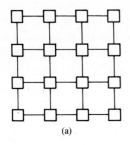

(a)

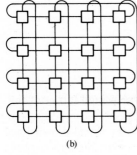

(b)

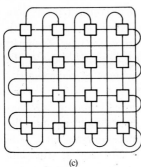

(c)

Figure 2-2 Two-dimensional meshes. (a) Mesh with no wrap-around connections. (b) Mesh with wrap-around connections between processors in same row or column. (c) Mesh with toroidal wrap-around connections.

has level number 0, and the single processor at the apex of the pyramid has level number $\log_4 p$. Every interior processor is connected to nine other processors: one parent, four mesh neighbors, and four children. Figure 2-3 illustrates a pyramid of size 16.

Shuffle-Exchange Network

A **shuffle-exchange network** consists of $n = 2^k$ nodes, numbered 0, 1, ..., $n - 1$, and two kinds of connections, called *shuffle* and *exchange*. Exchange connections link pairs of nodes whose numbers differ in their least significant bit. The **perfect shuffle** connection links node i with node $2i$ modulo $(n - 1)$, with the exception that node $n - 1$ is connected to itself. See Figure 2-4 for

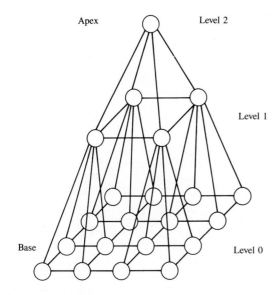

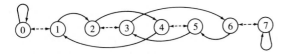

Figure 2-3 A pyramid network of size 16.

Figure 2-4 Shuffle-exchange network with eight nodes. Solid arrows denote shuffle connections. Dashed arrows denote exchange connections.

a drawing of an eight-node shuffle-exchange network. Shuffle connections are indicated by solid arrows; dashed arrows represent exchange links.

To understand the derivation of the name *perfect shuffle*, consider shuffling a deck of eight cards, numbered 0, 1, 2, 3, 4, 5, 6, 7. If the deck is divided into two exact halves and shuffled perfectly, then the result is the following order: 0, 4, 1, 5, 2, 6, 3, 7. Reexamine Figure 2-4: the final position of the card that began at index i can be determined by following the shuffle link from node i.

Let $a_k a_{k-1} a_{k-2} \cdots a_1$ be the address of a node in a perfect shuffle network, expressed in binary. A datum at this address will be at address $a_{k-1} a_{k-2} \cdots a_1 a_k$ following a shuffle operation. In other words, the change in the address of a piece of data after a shuffle operation corresponds to left cyclic rotation of the address by 1 bit (Exercise 2-1). If $n = 2^k$, then k shuffling operations move a datum back to its original location. The nodes through which a data item beginning at address i travels in response to a sequence of shuffles are called the **necklace** of i. No necklace is longer than $\log n$; a necklace shorter than $\log n$ is called a **short necklace**. Figure 2-5 illustrates the necklaces of the perfect shuffle network with eight nodes.

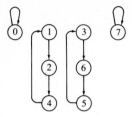

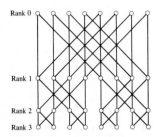

Figure 2-5 Necklaces of the shuffle-exchange network with eight nodes.

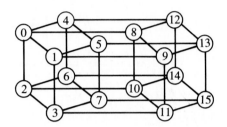

Figure 2-6 Butterfly network with 32 nodes. (Ullman [1984].)

Figure 2-7 Four-dimensional (16-node) hypercube.

Butterfly

A **butterfly network** consists of $(k + 1)2^k$ nodes divided into $k + 1$ rows, or *ranks*, containing $n = 2^k$ nodes each (Figure 2-6). The ranks are labeled 0 through k, although the ranks 0 and k are sometimes identified, giving each node four connections to other nodes.

Let node(i, j) refer to the jth node on the ith rank, where $0 \le i \le k$ and $0 \le j < n$. Then node(i, j) on rank $i > 0$ is connected to two nodes on rank $i - 1$: node$(i - 1, j)$ and node $(i - 1, m)$, where m is the integer found by inverting the ith most significant bit in the binary representation of j. Note that if node(i, j) is connected to node$(i - 1, m)$, then node(i, m) is connected to node$(i - 1, j)$. The entire network is made up of such "butterfly" patterns, hence the name. As the ranks decrease, the widths of the wings of the butterflies increase exponentially.

Hypercube (Cube-Connected)

A cube-connected network is a butterfly with its columns collapsed into single nodes. Formally, this network consists of $n = 2^k$ nodes forming a k-dimensional **hypercube**. The nodes are labeled 0, 1, ..., $2^k - 1$; two nodes are adjacent if their labels differ in exactly one bit position. A four-dimensional hypercube is shown in Figure 2-7.

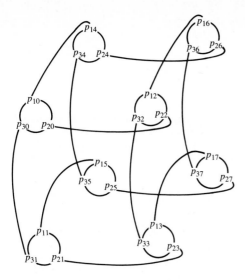

Figure 2-8 Cube-connected cycles network with 24 nodes. The first subscript of each node denotes the rank; the second subscript denotes the column. (Ullman [1984].)

Cube-Connected Cycles

The **cube-connected cycles network** is a k-dimensional hypercube whose 2^k "vertices" are actually cycles of k nodes formed by columns of a butterfly network whose ranks 0 and k have been identified. For every dimension, every cycle has a node connected to a node in the neighboring cycle in that dimension. See Figure 2-8 for a drawing of a 24-node cube-connected cycles network.

Formally, node(i, j) is connected to node(i, m) if and only if m is the result of inverting the ith most significant bit of the binary representation of j. Note that the connections are slightly different from those in the butterfly network: if node(i, j) is connected to node($i - 1$, m) in the butterfly network, where $j \neq m$, then node(i, j) is connected to node(i, m) in the cube-connected cycles network. However, node(i, j) can still communicate with node(i, m) by following two links, since there is a direct path from node(i, m) to node($i - 1$, m).

A parallel model is **reasonable** if the number of processors each processor can communicate with is bounded by a constant [Goldschlager 1982]. A natural question is whether each of these reasonable processor organizations (mesh, pyramid, butterfly, perfect shuffle, and cube-connected cycles) is superior for some class of problems or whether one of these is the best overall model. Galil and Paul [1981, 1983] show that a universal parallel model can simulate every reasonable parallel model "with only a small loss of time and with essentially the same number of processors." The heart of their universal computer is a sorting network that is used as a "post office" for sending and requesting information. Galil and Paul have shown that since cube-connected cycles are used as the sorting network, the cube-connected cycles network (and hence the butterfly) is an efficient general-purpose network.

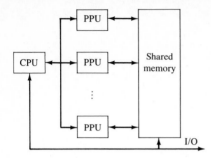

Figure 2-9 Diagram of a shared-memory SIMD model. The p processors are allowed to access simultaneously any p locations in the entire memory space in constant time.

2-3 PROCESSOR ARRAYS

A **processor array** (SIMD computer) is a vector computer implemented as a set of identical synchronized processing elements capable of simultaneously performing the same operation on different data. Although the processing elements execute in parallel, units may be programmed to ignore any particular instruction. This ability to mask processing elements allows synchronization to be maintained through the various paths of control structures, such as clauses of an if ... then ... else statement.

SIMD models can vary in two respects: the number of processors may be fixed or unbounded, and processors may communicate with each other via shared memory (SM), a mesh-connected (MC) network, a pyramid (P) network, a perfect shuffle (PS) network, a cube-connected (hypercube) (CC) network, or a cube-connected cycles (CCC) network.

Shared-Memory SIMD Model

The mechanism through which processors in an SIMD model access data has a significant impact on the time complexity of a parallel algorithm. Most reported parallel SIMD algorithms assume a shared global memory model that allows the p processors to access simultaneously any p locations in the entire memory space in constant time (Figure 2-9). The size of the shared memory is called the **communication width** [Vishkin and Wigderson to appear]. The shared-memory SIMD model consists of a control processor, a global random access memory, and a set of parallel processing units (PPUs) [Goldschlager 1982]. Each processor is a RAM, with the usual instruction set augmented by parallel instructions. The CPU (control processor) contains the machine's program and executes the serial instructions. It broadcasts the parallel instructions to the parallel processing units for execution. Each parallel processing unit has a unique index number, which allows different units to access different memory locations.

Numerous shared memory SIMD models have appeared in the literature. The model of Goldschlager [1982], called SIMDAG, allows multiple processors simultaneous read access to the same memory location. It also allows more than one processor to write to the same memory location simultaneously. In the case of simultaneous writing, the processor with the lowest index number takes precedence.

The P-RAM [Wyllie 1979] allows simultaneous read access, but does not allow simultaneous writing to the same location. The PP-RAM [Reif 1982] is a P-RAM with the added capability that processors are capable of doing independent probabilistic choices on a fixed input. Probabilistic parallel algorithms often have a much lower expected time complexity than nonprobabilistic parallel algorithms.

The SP-RAM [Shiloach and Vishkin 1982] allows simultaneous read access to the same memory location and simultaneous write access. In the case of simultaneous writing, no assumption is made about which processor succeeds in storing its value. The RP-RAM [Shiloach and Vishkin 1981] allows simultaneous reading and writing as well. If m processors attempt to write a value into the same memory location simultaneously, then exactly one processor succeeds, and the probability of each processor succeeding is $1/m$.

The EREW P-RAM (exclusive read, exclusive write P-RAM) [Snir 1982] forbids more than one processor from reading or writing the same memory cell simultaneously. The CREW P-RAM (concurrent read, exclusive write P-RAM) allows concurrent reads, but forbids concurrent writes. The CRCW P-RAM (concurrent read, concurrent write P-RAM) is a P-RAM that allows concurrent reads and concurrent writes.

Two kinds of shared-memory SIMD models appear later in this text: SIMD-SM and SIMD-SM-R. In the most restrictive SIMD-SM (shared-memory) model, no two processors may access the same memory location during the same instruction. Hence the SIMD-SM and EREW P-RAM models are identical. The SIMD-SM-R model allows any number of processors to address the same location for reading (accessing), but not for writing. The SIMD-SM-R model is identical to the P-RAM and CREW P-RAM models.

Despite their popularity, no actual processor arrays have been built based on the shared-memory model, because it is not feasible to allow p processors to access any p memory locations simultaneously. A more realistic assumption is that each processor has its own private memory, and processors can pass data only via a limited interconnection network (Figure 2-10) [Dekel, Nassimi, and Sahni 1981]. The following sections describe five SIMD models based on the mesh, pyramid, perfect shuffle, cube-connected, and cube-connected cycles networks.

Mesh-Connected SIMD Model

In the SIMD-MC^q (mesh-connected) model, the processors are arranged into a q-dimensional lattice. (See Figure 2-2 for illustrations of two-dimensional meshes.) The ILLIAC IV, Burroughs PEPE, Goodyear Aerospace MPP, and ICL DAP all fall into the category SIMD-MC^2.

The DAP. The DAP is a commercial parallel computer based upon the SIMD-MC^2 model. Bit-serial processors form a 64×64 processor array with nearest-

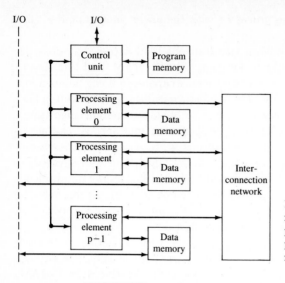

Figure 2-10 A more realistic SIMD model. Each processor has its own private memory, and processors can pass data only via a limited interconnection network.

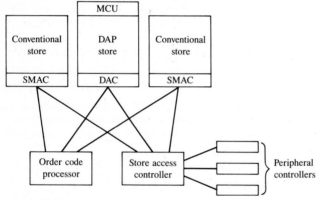

Figure 2-11 The DAP is integrated into an ICL 2900 series computer as a memory unit. (Hockney and Jessope [1981].)

neighbor connections. ICL in Great Britain manufactures the DAP; it delivered the first production models in 1980.

The DAP emulates a 2-megabyte memory module and has the additional ability to process data in a highly parallel manner. It is designed to be integrated into an ICL 2900 series computer. Figure 2-11 illustrates how one or more DAPs can be configured with an ICL 2900. The DAP provides memory and can also be instructed by the 2900's CPU to execute its own DAP program. The store access controller gives peripherals access to memory and transfers blocks of data between memory units. Thus in a typical program the order code processor provides serial pre- and postprocessing, while the DAP is assigned the task of doing the "number crunching" in parallel.

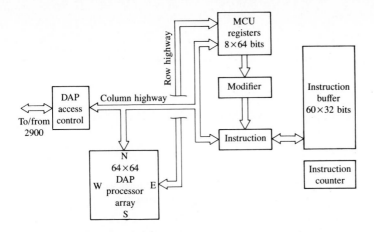

Figure 2-12 Major components of the DAP. (Hockney and Jesshope [1981].)

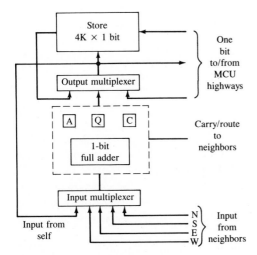

Figure 2-13 A DAP processing element. (Hockney and Jesshope [1981].)

The relationships between the major components of the DAP are shown in Figure 2-12; 4096 bit-serial processors are organized into a 64 × 64 array. Each 64-bit word in the 2900 corresponds to a row across the DAP memory. Orthogonal 64-bit row and column highways are capable of fetching arbitrary rows and columns from the memory. The column highway fetches data words from DAP memory, fetches rows to a master control unit register, and fetches pairs of 32-bit instructions, stored two per word. The instruction buffer can store up to 60 instructions, speeding the execution of small loops. The row highway transfers columns to and from master control unit registers.

A DAP processing element is diagramed in Figure 2-13. The processing elements are simple: one of the goals of the system architects was to use in-

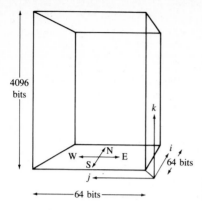

Figure 2-14 Memory address space of the DAP. (Hockney and Jesshope [1981].)

Table 2-1 Performance of the ICL DAP on various problems

Operation	Time (μs)	Processing rate (megaflops)
$Z \leftarrow X$	17	241
$Z \leftarrow X^2$	125	133
$Z \leftarrow X + Y$	150	27
$I \leftarrow MOD(I)$	1	4096
64×64 matrix multiplication	30	17

X, Y, and Z are 4096-element vectors containing 32-bit real numbers; I is a 4096-element vector containing 32-bit integers.

expensive, reliable technology to achieve high processing power. A processing element contains two multiplexers, three 1-bit registers, a 1-bit full adder, and a 4K × 1-bit memory. The A register provides programmable control (masking). The Q register is an accumulator, and the C register is for carry-save. And 4K × 1 bits of memory per processor multiplied by 4096 processors results in a total of 2 megabytes of memory. The memory address space is shown in Figure 2-14.

Table 2-1 gives the performance of the DAP on various problems. The DAP is best suited for problems requiring small word lengths.

Other SIMD Models

The SIMD-P (pyramid) model uses the pyramid network to organize processors. Parallel computers based on this model have been proposed for the purpose of performing low-level image processing. Pyramid models have the advantage, along with two-dimensional meshes, that a p-processor computer can be constructed in VLSI in $\Theta(p)$ area [Dyer 1981]. Several pyramid computers are under construction [Levialdi 1985; Tanimoto 1981].

The SIMD-CC (cube-connected) model uses the hypercube network to organize processors. The SIMD-CC model requires $\log p$ connections per processor.

For large p, this value is unsuitably high. The following two models are able to emulate the SIMD-CC model, while requiring only three connections per processor.

Processors in the SIMD-PS model are connected by using Stone's perfect shuffle network (Figure 2-4). Siegel [1979] has shown that a composition of k shuffle-exchange networks, called an **omega network**, is equivalent to a cube-connected network with degree k. The same effect can be achieved by building only one stage of the network and cycling through it k times [Lawrie 1975]. The SIMD-CCC model uses the cube-connected cycles network to link processors [Preparata and Vuillemin 1981].

Associative Processors

An **associative processor** is a special kind of processor array. Instead of having a random access memory, an associative processor is built around an associative memory that allows the simultaneous searching of the whole memory for specified contents. Sanders Associates' OMEN series of computers [Higbie 1972] and the Goodyear Aerospace STARAN [Batcher 1979] are examples of associative processors.

2-4 MULTIPROCESSORS AND MULTICOMPUTERS

Multiple-instruction stream multiple-data stream (MIMD) computers consist of a number of fully programmable processors, each capable of executing its own program. Multiprocessors are characterized by a shared memory. In contrast, every CPU in a multicomputer has its own private memory, and all communication and synchronization between processors must be via messages. Within these categories, further distinctions may be made by examining the processor interconnection pattern.

Tightly Coupled Multiprocessors

The simplest processor intercommunication pattern assumes that all the processors work through a central switching mechanism to reach a shared global memory (Figure 2-15). There are a variety of ways to implement this switching mechanism, including a common bus to global memory, a crossbar switch, and a packet-switched network. Carnegie-Mellon's C.mmp, Denelcor's HEP, Encore's Multimax, New York University's Ultracomputer, and Sequent's Balance 8000 are examples of tightly coupled multiprocessors.

In the case of systems using a bus, such as the Multimax or Balance 8000, it is impractical to build large systems of this type, because only so many processors can share a bus before the bus becomes saturated. In the case of systems using a crossbar switch, such as the C.mmp, the cost of the switch soon becomes the dominant factor, again limiting the number of processors which may be

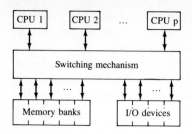

Figure 2-15 The tightly coupled multiprocessor model. All the processors work through a central switching mechanism to reach a shared global memory.

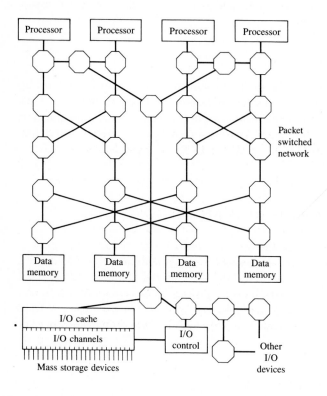

Figure 2-16 Architecture of a four-processor Denelcor HEP.

connected [Stone 1980]. Tightly coupled multiprocessors based on switching networks may contain a large number of processors. For example, New York University's Ultracomputer has 512 processors.

The HEP. The first commercially available tightly coupled multiprocessor was the HEP, manufactured by Denelcor of Denver, Colorado. It was designed to contain up to 16 process execution modules (PEMs) and up to 128 data memory modules, connected by a high-speed switching network. The architecture of a four-processor HEP is illustrated in Figure 2-16.

In order to achieve a consistently high processing rate, each PEM is capable of executing multiple instruction streams in a pipelined fashion. In other words, each PEM is capable of managing a large number of processes concurrently.

Figure 2-17 contrasts the processing rates achievable on SISD, SIMD, and MIMD processors. Even with instruction prefetching, the amount of concurrency achievable on an SISD processor (such as the CDC 6600) is limited, because of the existence of conditional branch instructions. Branches also limit the parallelism possible on SIMD machines, since only one processing element is needed to execute the branch. The remaining processing elements remain idle during branches and all scalar instructions. An MIMD computer, however, can keep multiple functional units busy, because each functional unit executes its own instruction stream. However, it is only fair to point out that, unlike the previous two methods, the last scheme does not achieve any speedup for a single instruction stream.

A PEM fetches an instruction every 100 nanoseconds. Control and data loops are pipelined into eight 100-nanosecond segments. At least 800 nanoseconds have to elapse between the start of two consecutive instructions in any process's instruction stream. This means that although a PEM can execute 10 million instructions per second (MIPS), it achieves this speed only when at least eight processes are being executed, and a single process can never execute faster than 1.25 MIPS.

Every PEM has its own local program memory, constant memory, and 2048 general-purpose registers, reducing the number of shared-memory accesses. Context switching is efficiently implemented in hardware.

The HEP uses a synchronous, pipelined, packet-switched network to transfer data between the PEMs and shared memory. Switching nodes receive message packets and attempt to reduce the distance between the packet and its destination. In case of a conflict for a path, the lower-priority packet is sent along an alternate route that may actually increase its distance from its destination. Each such "bump" increases the priority of the packet. The pipelining of the switches allows the pipelined PEMs to execute a large number of instruction streams concurrently.

Synchronization among cooperating processes is achieved through controlled access to shared data. Every register memory and data memory location on the HEP has an associated access state—*full* or *empty*. If so desired, an instruction trying to read a word from a memory cell can be prevented from executing until the location is full. Reading the location and setting its state to empty form an indivisible operation. Similarly, writing a value into a location can be prevented from succeeding when the state is empty, and writing and setting the state full form another indivisible operation.

Loosely Coupled Multiprocessors

Like tightly coupled multiprocessors, loosely coupled multiprocessors are characterized by a shared address space. Unlike tightly coupled multiprocessors, the

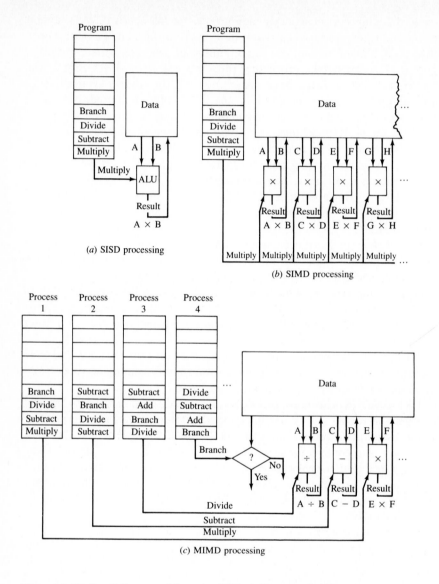

Figure 2-17 Parallelism possible under SISD, SIMD, and MIMD models.

shared address space on a loosely coupled multiprocessor is formed by combining the local memories of the CPUs. Hence the time needed to access a particular memory location on a loosely coupled multiprocessor depends on whether that location is local to the processor.

Cm*. The Cm* of Carnegie-Mellon University was the best-known example of a loosely coupled multiprocessor [Stone 1980]. The Cm* system was based on a

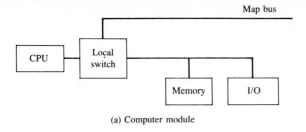

(a) Computer module

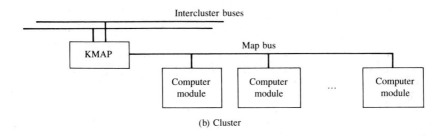

(b) Cluster

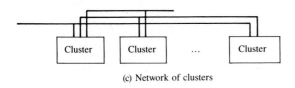

(c) Network of clusters

Figure 2-18 Architecture of the Cm*. (a) Computer module. (b) Cluster of computer modules. (c) Network of clusters.

computer module consisting of a processor, local memory, local I/O devices, and a local switch connecting the module to the rest of the system (Figure 2-18a). All processor references to I/O or memory went through the local switch. If an access was local, the reference was directed to the local memory or I/O device. Nonlocal references were directed to the *map bus*, the connection to the rest of the processors. Several computer modules were combined to form a cluster sharing a single map bus (Figure 2-18b). Clustering allowed processors to share data. Competition for the map bus limited the number of processors that could be connected into a single cluster. Clusters were connected via intercluster buses (Figure 2-18c). If the Kmap (map controller) detected a memory reference to another cluster, it redirected the reference via an intercluster bus to another Kmap, which placed the reference on its own map bus. Since loosely coupled

Processor node

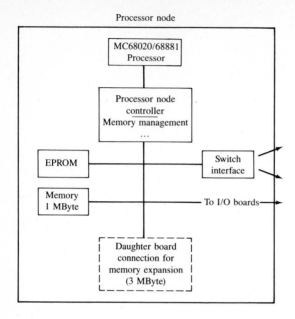

Figure 2-19 A processor node of the Bolt, Beranek, and Newman Butterfly. (Reprinted by permission of Bolt, Beranek, and Newman, Inc.)

multiprocessors do not have a centralized switching mechanism, a large number of processors may be connected. At one time Cm* had 50 active processors.

Bolt, Beranek, and Newman Butterfly™ Parallel Processor[†]. The Butterfly Parallel Processor consists of between 1 and 256 processor nodes, each of which contains a Motorola 68020 CPU, between 1 and 4 megabytes of primary memory, a coprocessor called the processor node controller (PNC), memory management hardware, an I/O bus, and an interface to the switching network (Figure 2-19).

The PNC has several important duties. First, every memory reference made by the CPU passes to the PNC. The PNC uses the memory management hardware to translate the virtual address used by the CPU into a physical memory address. If the reference is to a nonlocal memory address, the PNC initiates a reference to remote memory through the switching network. Likewise, the PNC handles references to the local memory initiated by PNCs at remote processor nodes. The name of the machine is derived from its use of a butterfly network to route remote memory accesses (Figure 2-20).

Another duty of the PNC is to provide implementations of operations needed for parallel processing, including test-and-set, queuing operations, event mechanisms, and a process scheduler that works with the queuing and event mechanisms to allow processes to communicate and synchronize with each other.

[†]*Butterfly*™ is a trademark of Bolt, Beranek, and Newman, Inc.

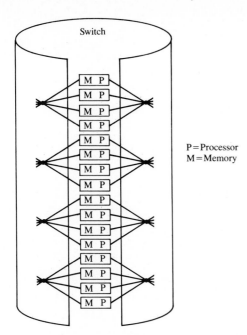

P = Processor
M = Memory

Figure 2-20 Nonlocal memory references on the Butterfly are sent through the switch. (Reprinted by permission of Bolt, Beranek, and Newman, Inc.)

Figure 2-21 illustrates the butterfly switch of the previous figure in more detail. Imagine that the cylinder of Figure 2-20 has been split down the side, between the processors and the memories, and then flattened. Each node of the switch is a VLSI switching element with four inputs and four outputs. Hence the number of switching elements in a p-processor machine is $p \log_4 p$. Data flow through the switch in packets; address bits route the data from the source PNC to the destination PNC.

Multicomputers

Another model of MIMD computation, the multicomputer, has no shared global memory. Each processor has its own local memory, and process cooperation occurs either through message passing or through memory shared between pairs of processors. Ametek's S/14, Intel's iPSC, and NCUBE's NCUBE/10 are commercial multicomputers.

Intel iPSC. The iPSC[†] is a commercial parallel computer marketed by Intel Scientific Computers of Beaverton, Oregon. The largest model consists of 128 microcomputer-based nodes connected in a binary 7-cube, also called a seven-dimensional hypercube or a cube-connected network of degree 7. A node contains

---

[†]*iPSC* is a trademark of Intel Corporation.

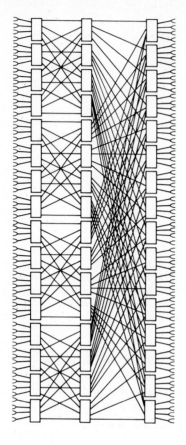

Figure 2-21 Switching network for a 64-processor butterfly. (Reprinted by permission of Bolt, Beranek, and Newman, Inc.)

an Intel 80286 central processing unit, an Intel 80287 floating-point coprocessor, and 512 kilobytes of dynamic RAM (Figure 2-22). Each node is connected to seven other nodes through fast bidirectional, asynchronous, point-to-point communication channels capable of transmitting 10 megabits per second. Each node has a copy of the operating system kernel that schedules and runs processes inside the node, provides system calls that allow processes to communicate with each other, and routes messages that pass through the node. Users interact with the iPSC through the "cube manager," an Intel System 286/310 microcomputer that has the ability to communicate with every processing node in the system over an Ethernet channel.

2-5 PARALLEL COMPUTING TERMINOLOGY

The **worst-case time complexity** (or simply **time** or **complexity**) of a parallel algorithm is a function $f(n)$ that is the maximum, over all inputs of size n, of the time elapsed from when the first processor begins execution of the al-

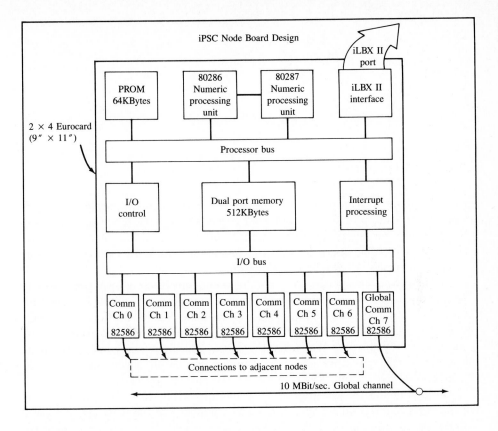

Figure 2-22 Node of Intel's iPSC. (Reprinted by permission of Intel Corporation.)

gorithm until the last processor terminates algorithm execution. For example, a sequential algorithm to determine the sum of n values has complexity $O(n)$, since it requires $n-1$ additions. If, however, additions are allowed to be done in parallel and $n/2$ processors are available, the sum can be determined in $\lceil \log n \rceil$ steps by computing partial sums in a treelike fashion (see Figure 2-23). Thus, parallel addition has complexity $O(\log n)$ with $n/2$ processors. Some authors use the term **depth** to refer to the complexity of a parallel algorithm. The **cost** of a parallel algorithm is defined as its complexity times p, the number of processors.

Two important measures of the quality of parallel algorithms implemented on multiprocessors and multicomputers are speedup and efficiency. The **speedup** achieved by a parallel algorithm running on p processors is the ratio between the time taken by that parallel computer executing the fastest serial algorithm and the time taken by the same parallel computer executing the parallel algorithm using p processors. The **efficiency** of a parallel algorithm running on p processors is the speedup divided by p.

An example illustrates the terminology. If the best known sequential algo-

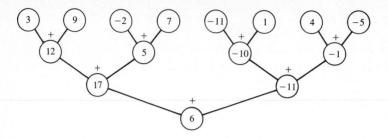

Figure 2-23 Determining sum of k values in parallel.

rithm executes in 8 seconds on a parallel computer, while a parallel algorithm solving the same problem executes in 2 seconds when five processors are used, then we say that the parallel algorithm "exhibits a speedup of 4 with five processors." A parallel algorithm that exhibits a speedup of 4 with five processors "has an efficiency of 0.8 with five processors."

Some define the speedup of a parallel algorithm running on p processors to be the time taken by the parallel algorithm on one processor divided by the time taken by the parallel algorithm using p processors. This definition can be misleading. Parallel algorithms frequently contain extra operations to accommodate parallelism. Comparing the speedup of a parallel algorithm running on many processors with that same algorithm running on one processor exaggerates the speedup, because it hides the overhead of the parallel algorithm.

Others define the speedup of a parallel algorithm on a particular parallel computer to be the ratio between the execution time of the most efficient serial algorithm running on the fastest serial computer and the execution time of the parallel algorithm running on the parallel computer. Although this definition may in fact be the fairest way to compare parallel algorithms with serial algorithms, its adoption would have the practical consequence of making it virtually impossible to make a claim about speedup, since few researchers have access to the most powerful serial computers, and benchmarks are not usually available.

Can Speedup Be Greater than Linear?

A parallel algorithm is said to exhibit **linear speedup** if the speedup with p processors is $\Theta(p)$. Is superlinear speedup possible? That is, is it possible for the speedup achieved by a parallel algorithm to be greater than the number of processors used? The answer depends upon your point of view. Let us examine both sides of the argument.

Superlinear speedup is impossible. Speedup cannot be greater than linear, because a sequential computer can always emulate a parallel computer. Suppose a parallel algorithm A_p solves an instance of problem Π in T_p units of time on a parallel computer with p processors. Then algorithm A_p can solve the same problem instance in $p \times T_p$ units of time on the same computer with one processor through time slicing. Hence the speedup cannot be greater than p. Because parallel algorithms usually have an associated overhead, most likely there is a sequential algorithm that solves the problem instance in time less than $p \times T_p$, and then the speedup would be less than linear.

Superlinear speedup is possible. Speedup can be greater than linear, because it is unfair to choose the algorithm after the problem *instance* is chosen. Speedup is supposed to measure the time taken by the best sequential algorithm divided by the time taken by the parallel algorithm, but it is going too far to allow the definition of *best* to change every time the problem instance changes. In other words, fairness dictates that the best sequential algorithm and the parallel algorithm be chosen before the particular problem instance be chosen. In this case it is possible for the parallel algorithm to exhibit superlinear speedup, because the extra processors may enable it to "luck out" and come upon the solution very quickly.

In this text we adopt the latter point of view, and we see examples of superlinear speedup later in the book.

2-6 AMDAHL'S LAW REVISITED

What is the best strategy for building high-performance parallel computers? The surest route to success, say some, is to connect a relatively small number of extremely powerful processors. Let us call this the "herd of elephants" approach. At the opposite extreme, others advocate connecting massive numbers of less powerful processing elements, taking the "army of ants" approach, so to speak. What can be said about these alternatives?

Grosch's law is true within classes of computers (mainframes, minicomputers, microcomputers), but it is not true between classes. Within any class, such as microcomputers, the power of the computer rises more quickly than its cost, but you can buy more megaflops per dollar by spending your money on top-of-the-line microcomputers rather than on pipelined vector processors [Ein-Dor 1985]. In other words, if you have a fixed budget to spend on a supercomputer, you can buy the most raw megaflops by spending your money on expensive microprocessors. Why not buy buckets full of the most sophisticated microprocessors available and wire them together?

Recall Amdahl's law: If f is the inherently sequential fraction of a computation to be solved by p processors, then the speedup S is limited according to the formula

$$S \leq \frac{1}{f + (1 - f)/p}$$

Suppose that a parallel computer E has been built by using the herd-of-elephants approach, where each processor in the computer is capable of performing sequential operations at the rate of X megaflops. Suppose that another parallel

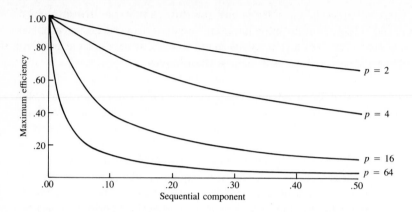

Figure 2-24 Maximum efficiency as a function of sequential component.

computer A has been built by using the army-of-ants approach; each processor in this computer is capable of performing sequential operations at the rate of αX megaflops, for some α where $0 < \alpha \ll 1$. If parallel computer A attempts a computation whose inherently sequential fraction f is greater than α, then A will execute the algorithm more slowly than a single processor of parallel computer E. (Why?) Hence a sequential component that may be acceptable for an algorithm executing on computer E may be unacceptable for an algorithm executing on computer A.

It does no good to have a lot of processor power that goes to waste; processors must maintain some level of efficiency. As the number of processors in a parallel computer increases, it becomes more and more difficult to use these processors efficiently. For example, let us determine what portion of a computation can be sequential if we want the processors to be used at 50 percent efficiency. Saying that p processors exhibit an efficiency of 50 percent is the same as saying that the speedup achieved is $p/2$. Hence

$$\frac{p}{2} \leq \frac{1}{f + (1 - f)/p}$$

$$\Rightarrow \quad pf + 1 - f \leq 2$$

$$\Rightarrow \quad f \leq \frac{1}{p - 1}$$

In other words, in order to maintain a constant efficiency, the fraction of the computation devoted to sequential processing must be proportional to the inverse of the number of processors. Figure 2-24 illustrates the maximum efficiency achievable by various size parallel computers for different values of f.

It is very important that every activity be considered when the sequential component is determined. For example, if a computer cannot tolerate much sequential processing, then I/O must be done in parallel as well. If parallelism eliminates the bottleneck in the CPU only to expose a bottleneck in the I/O unit, then the architecture is flawed. Computers capable of high processing speeds must have high memory bandwidth and high I/O rates, too.

If a parallel computer built by using the army-of-ants approach is to be successful, one of the following conditions must be true: at least one processor must be capable of extremely fast sequential computations, or the problem being solved must admit a solution containing virtually no sequential component.

2-7 SUMMARY

This chapter has discussed six processor organizations and several parallel models of computing. Processor organizations that have been suggested for parallel computers include the mesh, pyramid, perfect shuffle, butterfly, hypercube, and cube-connected cycles. The butterfly, hypercube, and cube-connected cycles organizations are closely related.

The random access machine (RAM) model of serial computation is important to students of parallel algorithms. Each element of a processor array can be thought of as a random access machine augmented with a few parallel instructions. Processor array (SIMD) models are characterized by how the processors interact with memory. Shared-memory SIMD models assume a shared global memory that allows processors access to arbitrary memory locations. Other SIMD models assume that each processing element has its own local memory and that information is transferred from one processing element to another through one of the processor organizations described above. Processor arrays are usually attached to a host computer.

MIMD computers are a more general-purpose model of parallel computation, since they consist of a number of CPUs asynchronously executing independent instruction streams. MIMD computers can be categorized by how the CPUs access memory. Tightly coupled multiprocessors have a single, shared address space, and the distance from a CPU to any memory location is a constant. Loosely coupled multiprocessors also have a shared address space, but each memory cell is closer to one CPU than to the others. Multicomputers have no shared memory. Each CPU has its own private address space, and processors interact through message passing.

Speedup and efficiency are two important measures of the quality of a particular parallel algorithm implementation. Given the best sequential algorithm and a parallel algorithm, there may exist particular instances for which the parallel algorithm exhibits superlinear speedup (and hence efficiency > 1) for certain numbers of processors.

Some people have advocated building parallel computers out of massive numbers of primitive processing elements. These parallel computers will be able to compete with other supercomputers only if they have at least one processor capable of extremely fast sequential operation or if they execute algorithms with virtually no sequential component.

BIBLIOGRAPHIC NOTES

Aho, Hopcroft, and Ullman [1974] describe the random access machine in greater detail.

The texts by Siegel [1984], Ullman [1984], and Uhr [1984] contain a wealth of information on processor interconnection methodologies. Feng [1981] also surveys connection methods. Miller and Stout [1986] have written an authoritative paper on pyramid computers. The perfect shuffle network was introduced by Stone [1971]. Additional references on shuffle-exchange networks include those by Lang and Stone [1976] and Wu and Feng [1981]. Lawrie [1975] introduced the omega network, a multistage network based on perfect shuffle connections. Siegel [1979] shows how other interconnection networks can simulate the perfect shuffle.

Two other processor organizations, not discussed in this chapter, deserve mention. Rosenfeld [1985] has proposed the prism network as an alternative to the pyramid. A prism network contains as many levels as a pyramid, but, unlike a pyramid, every level contains the same number of processors. Prisms have a number of interesting attributes. A prism can simulate a pyramid in linear time. In addition, a prism can compute the fast Fourier transform in linear time. The mesh-of-trees network has captured the interest of some theoreticians. Properties of this topology are discussed in Ullman [1984].

A large number of references exist for processor arrays. Books by Thurber [1976], Kuck [1978], Stone [1980], and Hwang and Briggs [1984] introduce the subject. Paul [1978], Thurber [1979], and Hwang, Su, and Ni [1981] have compared various processor array architectures. Hockney and Jesshope [1981] cover the ICL DAP in their monograph. The DAP is also discussed by Reddaway [1979]. Barnes et al. [1968] and Falk [1976] are references for the ILLIAC IV computer. Batcher [1980] and Fung [1977] describe Goodyear's Massively Parallel Processor. Crane et al. [1972] describe PEPE.

Yau and Fung [1977] have surveyed associative processor architectures. In addition, associative processors are described in texts by Thurber [1976] and Foster [1976].

Another perspective on multiprocessor architectures can be found in Enslow [1977, 1978].

Fuller and Oleinick [1976] and Mashburn [1979] have written about C.mmp, an early tightly coupled multiprocessor developed at Carnegie-Mellon University. A more complete description of the Denelcor HEP is in Hwang and Briggs [1984]. Kowalik [1985] has assembled a variety of articles on the HEP. An early description of the HEP was written by Smith [1978]. Gottlieb [1986] and Gottlieb et al. [1983] describe the New York University Ultracomputer, a modern tightly coupled multiprocessor.

Stone [1980] and Swan et al. [1977] are two of many sources of information on Cm*. See BBN [1985] for more information on the Butterfly. Seitz [1985] describes the implementation of the Cosmic Cube, a precursor of the iPSC.

EXERCISES

2-1 Given a perfect shuffle network, prove that if a shuffle link connects nodes i and j, then j is a single-bit left cyclic rotation of i.

*2-2 Prove that the number of necklaces in an n-node shuffle-exchange network is $O(2^k/k)$.

2-3 Name two ways to implement vector computers.

2-4 Explain why the ICL DAP is best suited for problems requiring small word lengths.

2-5 The Burroughs Scientific Processor, a processor array, was designed to have 16 arithmetic units and 17 memory banks. Come up with a reason for having the number of memory banks relatively prime to the number of arithmetic units.

2-6 What are the advantages and disadvantages of moving from the tightly coupled multiprocessor model to the loosely coupled multiprocessor model?

2-7 What is the difference between a binary k-cube and a cube-connected network of degree k?

2-8 Show that parallel addition has complexity $O(\log n)$ with $n/\log n$ processors.

2-9 Is it possible for the average speedup exhibited by a parallel algorithm to be superlinear?

2-10 Suppose a multiprocessor is built out of individual processors capable of sustaining 0.5 megaflop. What is the largest fraction of a program's execution that could be devoted to sequential operations if the parallel computer is to exceed the performance of a supercomputer capable of sustaining 50 megaflops?

THREE

DESIGNING PARALLEL ALGORITHMS

There are at least three ways to design a parallel algorithm to solve a problem. You can detect and exploit any inherent parallelism in an existing sequential algorithm, you can invent a new parallel algorithm, or you can adapt another parallel algorithm that solves a similar problem. Each method has its place. Unless you are the first person to try to solve a particular problem by using a computer, a sequential algorithm already exists. Rather than reinventing the wheel, so to speak, you may be able to profit from the work done by others. Perhaps the sequential algorithm can be made parallel in a straightforward manner.

However, blindly transforming a sequential algorithm to parallel form is often a mistake. Some sequential algorithms have no obvious parallelization; a parallel algorithm made from such a sequential algorithm will exhibit poor speedup. The architecture, too, often demands that a new approach be taken. Thus, sometimes it is better to "start from scratch."

Studying parallel algorithms can result in the development of general techniques that apply in a wide variety of circumstances. Hence it is often possible to adapt a successful parallel algorithm to solve a similar problem.

Section 3-1 presents a few general statements about the process of designing a parallel algorithm. Section 3-2 describes the development of parallel algorithms for processor arrays. Because most processor array models differ only in the way in which the processors are connected, communication issues are particularly important. The importance of communication issues is illustrated with a running example.

Sections 3-3 through 3-6 discuss topics related to algorithm development on multiprocessors and multicomputers. Section 3-3 describes three kinds of

MIMD algorithm: pipelined algorithms, relaxed algorithms, and partitioned algorithms. Factors that can limit the speedup of algorithms running under this model are examined. Section 3-4 is concerned with the mechanisms by which parallel processes can interact as well as how these mechanisms are expressed in programming languages. Deadlock can occur whenever multiple processes share resources in an unsupervised manner; handling deadlock is the topic of Section 3-5. Section 3-6 addresses the problem of allocating tasks to processors.

3-1 GENERALITIES

Insight Has an Important Role

Suppose you are trying to come up with a parallel algorithm to solve a particular problem Π. If there is a well-known sequential algorithm to solve Π, you may choose to use it as a starting point. If the sequential algorithm is not particularly parallelizable, then you must be able to apply some external knowledge of the problem in order to break it up.

For example, consider the problem of finding the sum of n integer values, where $n > 0$. Undoubtedly a sequential algorithm to solve this problem springs to mind. Perhaps the algorithm has this form:

```
SUMMATION (SISD):
begin
    sum ← a₀
    for i ← 1 to n − 1 do
        sum ← sum + aᵢ
    endfor
end
```

Suppose $n = 4$. Then the additions would be done in this order:

$$[(a_0 + a_1) + a_2] + a_3$$

Is this an inherently sequential process? Given the way the parentheses and brackets are grouped, the sum of a_0 and a_1 must be found before a_2 can be added to the subtotal. Likewise, a_3 cannot be added to the sum until a_2 has been added to the sum $a_0 + a_1$. Of course, we know that the addition can be done in parallel. We have some external knowledge: we know that integer addition is associative on digital computers (well, almost). Because of associativity, we could write this expression as follows:

$$(a_0 + a_1) + (a_2 + a_3)$$

This grouping makes clear that while the first two elements are being added, the second pair can be added as well.

Communication Costs Must Be Considered

It is a mistake to ignore communication costs in determining the complexity of a parallel algorithm. Sometimes the communication complexity is higher than the computational complexity; in other words, more time is spent routing data among processors than actually manipulating the data.

For example, suppose it takes i units of time to perform a floating-point addition on a particular multicomputer and $100i$ units of time to pass a floating-point number from one processor to another. Assume that we must add n floating-point numbers on a p-processor system, and assume that these n numbers begin in the local memory of whatever processor(s) we desire. If $n \leq 101$, then no matter how large p is, it is faster to add the numbers on a single processor than to perform the additions in parallel, since all 100 additions can be done in the time of a single communication.

The Algorithm Must Fit the Architecture

The performance of an algorithm can be radically different on different architectures. Sometimes this is due to communication overhead, as explained above. Sometimes other factors are involved. For example, synchronization is automatic on the processor array model; the processors always work in lockstep. On multiprocessors and multicomputers, however, synchronization must be achieved through software and hence is time-consuming. Algorithms that perform relatively few operations between synchronizations (i.e., have a small **grain size**) usually exhibit poor efficiency on MIMD computers.

3-2 DEVELOPING ALGORITHMS FOR PROCESSOR ARRAYS

The processing elements of processor arrays work in lockstep. The algorithm designer does not have to worry about synchronizing the processors—that function is built into the architecture. Communication costs are not zero, however, and algorithms must be carefully planned to minimize the amount of communication needed. Sometimes the communication complexity dominates the computational complexity.

To illustrate, let us examine algorithms to find the sum of n values on the SIMD-CC (cube-connected) model, the SIMD-PS (perfect shuffle) model, and the SIMD-MC^2 (mesh-connected with dimension 2) model. In each case assume that $n = 2^m = l^2$, where m and l are positive integers and the n values to be added, denoted $A = (a_0, a_1, \ldots, a_{n-1})$, are distributed one value per processing element.

This is a good time to introduce some terminology that is used in the algorithms in the remainder of the book. Recall that p is always assumed to be the number of processors. Processing elements will be denoted by a P, with one

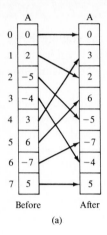 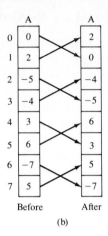

Figure 3-1 Two important procedures defined for the SIMD-PS model: (a) shuffle; (b) exchange.

or more subscripts being used to distinguish between processors. Parallelism is indicated by the use of the **for all** statement. Unlike a conventional **for** loop, the **for all** statement activates a set of processors that execute the statements before the matching **endfor** in parallel. In the case of SIMD computers, the processing elements execute the statements in lockstep; MIMD algorithms must assume that the CPUs are executing the statements asynchronously. In SIMD models each processor has the ability to read the value of a particular memory location in an adjacent processor. The symbol $\Leftarrow$ denotes such a communication of a data item from an adjacent processor's local memory into the "active" processor's local memory. Two procedures are important on the SIMD-PS model: **shuffle** and **exchange**. These procedures are illustrated in Figure 3-1.

The following algorithm adds $n = 2^m$ values on the SIMD-CC model. Processor P_i possesses local variables a_i and t_i, for all i such that $0 \le i \le n-1$. When the algorithm begins execution, the a_i's contains the values to be added. At its termination a_0 contains the sum.

```
SUMMATION (SIMD-CC):
begin
    for i ← log n − 1 downto 0 do
        d ← 2^i
        for all P_j, where 0 ≤ j < d do
            t_j ⇐ a_{j+d}
            a_j ← a_j + t_j
        endfor
    endfor
end
```

The outer **for** loop iterates $\log n$ times. Every loop iteration requires constant

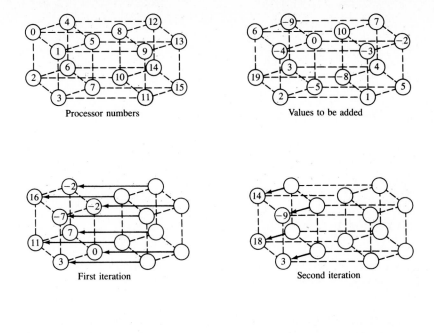

Processor numbers   Values to be added

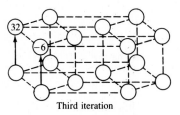
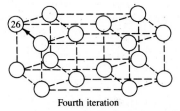

First iteration   Second iteration

Third iteration   Fourth iteration

Figure 3-2 Finding sum of 16 values on the SIMD-CC model.

time. Hence the complexity of this algorithm is $\Theta(\log n)$. The algorithm is illustrated in Figure 3-2 for $n = 16$.

The next sum-finding algorithm adds $n = 2^m$ values in $\log n$ steps on the SIMD-PS model.

```
SUMMATION (SIMD-PS):
begin
    for i ← 1 to log n do
        for all Pⱼ, where 0 ≤ j < n do
            shuffle (aⱼ)
            bⱼ ← aⱼ
            exchange (bⱼ)
            aⱼ ← aⱼ + bⱼ
        endfor
    endfor
end
```

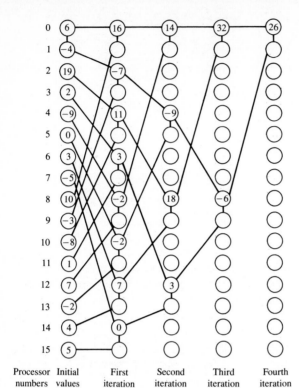

Processor numbers	Initial values	First iteration	Second iteration	Third iteration	Fourth iteration

Figure 3-3 Finding sum of 16 values on the SIMD-PS model.

Notice that the algorithm for the SIMD-PS model adds the same numbers in the same order as the algorithm for the SIMD-CC model. Once again, the value of a_0 contains the sum when the algorithm terminates. (What other processors contain the sum when the algorithm terminates?) However, the processor connection pattern on the SIMD-PS model is different. That is why the algorithm had to be modified slightly. In the case of this particular algorithm, every data routing required on the SIMD-CC model can be performed by two data routings—a shuffle followed by an exchange—on the SIMD-PS model. Since there are $\log n$ iterations and every iteration takes constant time, this parallel algorithm, too, has complexity $\Theta(\log n)$. Figure 3-3 illustrates the action of this algorithm as it adds 16 values.

It is easy to find the sum on the SIMD-CC and SIMD-PS models, because both models enable values in processors whose indices differ by exactly 1 bit to be brought together efficiently. Implementing the algorithm on the SIMD-MC^2 model is a bit more involved, because the SIMD-MC^2 model does not have this capability. One parallel algorithm to do the task is presented below. Recall that $n = l^2$. For simplicity, the n values to be added are stored, one per processing element, in locations designated $a_{1,1}, a_{1,2}, \ldots, a_{1,l}, a_{2,1}, \ldots, a_{l,l}$. The algorithm works by summing all the rows in column 1 and then summing column 1. When the algorithm completes, element $a_{1,1}$ contains the sum.

```
SUMMATION (SIMD-MC²):
begin
    for i ← l − 1 downto 1 do
        for all P_{j,i}, where 1 ≤ j ≤ l do  {Column i active}
            t_{j,i} ⇐ a_{j,i+1}
            a_{j,i} ← a_{j,i} + t_{j,i}
        endfor
    endfor
    for i ← l − 1 downto 1 do
        for all P_{i,1} do  {Only a single processing element active}
            t_{i,1} ⇐ a_{i+1,1}
            a_{i,1} ← a_{i,1} + t_{i,1}
        endfor
    endfor
end
```

The workings of this algorithm on an example can be found in Figure 3-4. It requires $2(l - 1) = 2\sqrt{n} - 2$ constant-time iterations; hence its complexity is $\Theta(\sqrt{n})$. Is this an optimal algorithm for the SIMD-MC^2 model? Yes, it is optimal, because $2\sqrt{n} - 2$ data routings are necessary to bring together $a_{1,1}$ and $a_{l,l}$, and the sum requires these two values, along with all the rest. Hence the SIMD-MC^2 model cannot add a series of numbers as quickly as the SIMD-PS and SIMD-CC models: the inefficiency of the interprocessor communication pattern **prevents** the numbers from being added any faster.

3-3 DEVELOPING ALGORITHMS FOR MIMD COMPUTERS

This section and the three that follow address four fundamental questions for the programmer of multiprocessors or multicomputers. First, given a problem with a certain amount of parallelism, how do you divide work among a number of processes so that they can work efficiently toward a solution? Second, what mechanisms allow processes to work together, and how are these mechanisms expressed in a programming language? Third, what is deadlock, and how can it be avoided? Fourth, how are processes scheduled on processors?

The remainder of this section is devoted to answering the first question: How do you divide work among a number of processes? The designer of parallel algorithms for MIMD computers has the same goal as the designer of algorithms for processor arrays: Given a problem with a certain amount of inherent parallelism and a number of processors, find an algorithm that best utilizes these processors to solve the problem as quickly as possible. However, MIMD computers are more general-purpose machines, since they enable the asynchronous execution of multiple instruction streams. Thus the designer of algorithms for MIMD computers has more flexibility.

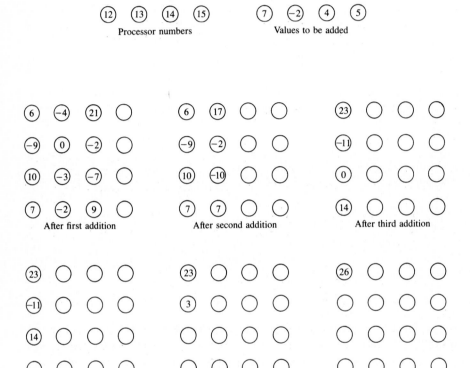

Figure 3-4 Finding sum of 16 values on the SIMD-MC^2 model.

Categorization of MIMD Algorithms

Parallelization on MIMD models can be divided into three categories: pipelined algorithms, partitioned algorithms, and relaxed algorithms.

Pipelined Algorithms. A **pipelined algorithm** is an ordered set of segments in which the output of each segment is the input of its successor. The input to the algorithm serves as the input to the first segment; the output of the last segment is the output of the algorithm. As in all pipelines, all segments must produce results at the same rate, or else the slowest segment will become a bottleneck. Some authors use the term **macropipelining** to refer to this kind of algorithm.

A parallel compiler whose individual phases—scanning, parsing, code generation, and code optimization—were assigned to sets of processors would be an example of a pipelined algorithm. The speedup achieved overall would be limited to the minimum speedup achieved by the four segments.

A **systolic algorithm** is a special kind of pipelined algorithm. Three attributes distinguish systolic algorithms. First, the flow of data is rhythmic and regular. Second, data can flow in more than one direction. Third, the computations performed at each segment are essentially identical. Like a pipelined algorithm, a systolic algorithm requires implicit synchronizations between processes producing data and processes consuming data.

Partitioning.   Unlike pipelining, in which processors assume different computational duties, **partitioning** is the sharing of a computation. A problem is divided into subproblems that are solved by individual processors. The solutions to the subproblems are then combined to form the problem solution. This pooling of solutions implies synchronization among the processors; for this reason partitioned algorithms are sometimes called **synchronized algorithms**.

To illustrate partitioning, consider the problem of adding kp values on a multiprocessor with p processors. A natural way to accomplish the summation is to initialize to 0 the value of a global variable that will eventually contain the grand total and then to create p processes, one per processor. Each process adds a unique set of k values. When a process has computed its subtotal, it adds the subtotal to the value of the grand total. The process must have exclusive access to the global variable while updating its value. After adding its subtotal to the grand total, the process terminates. After all p processes have terminated, the global variable contains the grand total.

Partitioned algorithms can be divided into two categories: prescheduled algorithms and self-scheduled algorithms. In a **prescheduled algorithm** each process is allocated its share of the computation at compile time. The algorithm described above is prescheduled, because it encodes the decision that each process should add an equal number of values. Prescheduling is the norm when a computation consists of a large number of subtasks, each of a known complexity.

Sometimes, however, a computation can be broken into a number of subtasks, but the exact amount of time needed to perform its subtasks is highly variable. Two common techniques are used in this situation. The first technique is to use prescheduling, but to make sure that there are far more subtasks than processes. If each process executes a large number of subtasks and this set of subtasks is a random selection from the entire set of subtasks, then there is a good probability that every process will end up doing about the same amount of work. This method has been shown effective [Weide 1981].

The second technique is to use **self-scheduling**, in which the work is not assigned to the processes until run time. In a self-scheduled algorithm a global list of work to be done is kept, and when a process is without work, another task is removed from the list. Processes schedule themselves as the program executes, hence the name.

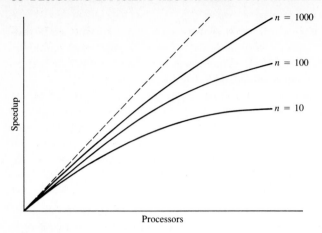

Figure 3-5 The Amdahl effect. As a general rule, speedup is a nondecreasing function of the problem size.

Relaxation. An algorithm that works without process synchronizations is said to be **relaxed**. All the processes may be working toward the same goal (as in partitioning), or there may be some specialization of purpose (as in pipelining), but the essential element is that no processor ever has to wait for another processor to provide it with data. Instead, relaxed algorithms are characterized by the ability of processors to work with the most recently available data. It is this nondeterministic behavior of algorithms parallelized by relaxation that makes prediction of performance difficult [Stone 1973]. Some authors prefer the term **asynchronous algorithm** to relaxed algorithm.

Summary. Parallel algorithms on MIMD models fall into one of three categories: pipelined algorithms, partitioned algorithms, or relaxed algorithms. In reality the distinctions often become blurred, and a parallel algorithm may have features of all three types. For example, an algorithm may be divided into segments at the highest level, forming a pipeline. One or more of these segments may be parallelized further via partitioning, while other segments may be parallelized via relaxation.

Factors That Limit Speedup

A number of factors can contribute to limit the speedup achievable by a parallel algorithm on an MIMD model. An obvious constraint is the size of the input problem. If there is not enough work to be done by the number of processors available, then any parallel algorithm would show constrained speedup. This phenomenon is called the **Amdahl effect** [Goodman and Hedetniemi 1977], and it explains why speedup is almost universally an increasing function of the size of the input problem (Figure 3-5).

The number of process creations and synchronizations in a partitioned algorithm must be minimized. Since process creations are more expensive than process synchronizations, the usual tactic is to create the desired number of processes when the algorithm begins execution and to synchronize them when necessary. If synchronizations must be done frequently, then the overhead can be significant. Because synchronizations are so expensive, a goal of the parallel algorithm designer should be to make the **grain size**—relative amount of work done between synchronizations—as large as possible, while keeping all the processors busy.

Sequential code limits the speedup of all three kinds of parallel algorithms (Amdahl's law). There is a more subtle limitation caused by the use of sequential code in partitioned algorithms. If a portion of the algorithm must be executed sequentially by one of the p processors, then the remaining $p - 1$ processors must wait for the sequential portion to complete before they resume. This implies synchronization among the processors.

The possible number of processors that can be put to work on the algorithm should not be too constrained, particularly by data structures that do not distribute processor contention. Too much contention for a single resource limits the speedup of a pipelined, partitioned, or relaxed algorithm in much the same way as the presence of sequential code does. Contention between processors for data resources is called **software lockout**.

A similar problem can occur at a more fundamental level. If the instructions used by all the processors in a tightly coupled multiprocessor are kept in a single shared memory bank, then processor contention for that memory bank puts a low ceiling on the speedup achievable by any parallel algorithm. An experiment by Oleinick at Carnegie-Mellon University dramatically illustrated the disastrous effects of keeping all the code in a single shared memory bank. See Figure 3-6. The lines represent the times required by parallel algorithms to find the root of a function, given various numbers of processors. The dotted line represents an implementation in which the code for every processor was stored in the same page of memory. Note processor contention for the single page allows hardly any speedup to occur. In fact, seven processors took longer to find the root than a single processor. The solid line represents an implementation in which the code for each processor was put on a unique memory page.

The ideal situation is for every processor in a multiprocessor to have a local memory where instructions, local variables, and constants are stored. "Constants" include global variables that are read but never written. That way shared memory is accessed only to find the value of a variable that is (1) shared among processes and (2) subject to modification. In other words, shared memory is used only for communication and synchronization between processes.

The workload must be balanced among the processors. Typically, workload balancing is not a problem for partitioned algorithms running on tightly coupled multiprocessors. In general, though, the task allocation problem is critical. There are two general allocation policies: **Static decomposition** assumes that the tasks and their precedence relations are known before execution, while **dynamic decomposition** assumes that tasks are generated during program ex-

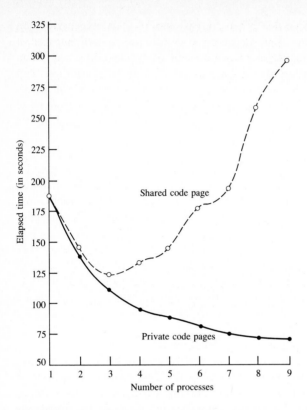

Figure 3-6 Processor contention for a single memory bank containing program code severely constrains speedup. Giving each processor a private code page eliminates this contention.

ecution. The advantage of static decomposition is that it allows the preallocation of tasks to processors. This can reduce the amount of interprocessor communication, which can be especially important on multicomputers. The advantage of dynamic decomposition is that it makes it easier to keep all the processors busy.

Finally, as efficient an algorithm as possible must be used. It is generally acknowledged that the fastest serial algorithms do not necessarily make the most efficient parallel algorithms [Baer 1982; Hockney and Jesshope 1981; Stone 1973]. However, since speedup is defined to be the ratio between the execution time of the best sequential algorithm and the time of the parallel algorithm, it is important that as little time complexity as possible be added when an algorithm to parallelize is chosen.

For example, assume that we are trying to solve problem Π on a parallel computer. Suppose that the time needed by the fastest sequential algorithm A_1 solving Π is $20n^2$ time units, when it is running on the parallel computer using a single processor. Suppose algorithm A_1 is not amenable to parallelization, and we choose to parallelize algorithm A_2, whose running time is $10n^{2.5}$. How much speedup can we expect to achieve by using our parallel version of A_2, given a problem of size $n = 100$? Since $20(100^2) = 200{,}000$ and $10(100^{2.5}) = 1{,}000{,}000$, the speedup of the parallel algorithm is 0.2 with one processor. It will take many processors to achieve a reasonable speedup over sequential algorithm A_1. (How many processors are needed to achieve a speedup of 1?)

Analyzing the Complexity of an MIMD Algorithm. Consider the following partitioned algorithm to find the sum of n values on a tightly coupled multiprocessor with p processors. The processors are labeled $P_0, P_1, \ldots, P_{p-1}$. In global memory are variables $a_0, a_1, \ldots, a_{n-1}$, containing the values to be added. Global variable *global_sum* is where the result is to be stored. Each processor P_i has local variable j_i, which is used as a for loop index, and *local_sum$_i$*, which contains the processor's subtotal. Procedures lock and unlock implement a critical section; locking is an atomic action, and no process can lock a locked variable. The for all statement is executed by p processes. The processes asynchronously execute the statements in the for all structure and terminate at the endfor. Once every process has reached the endfor, a single process resumes execution with the next statement after the endfor. (In this particular example there are no statements after the endfor.) Each process has local variables *local_sum* and j.

```
SUMMATION (TIGHTLY COUPLED MULTIPROCESSOR):
begin
    global_sum ← 0
    for all Pᵢ, where 0 ≤ i ≤ p − 1 do
        local_sum ← 0
        for j ← i to n step p do
            local_sum ← local_sum + aⱼ
        endfor
        lock (global_sum)
            global_sum ← global_sum + local_sum
        unlock (global_sum)
    endfor
end
```

Let us determine the worst-case time complexity of this parallel algorithm. First, how long does it take to create the processes? If the initial process creates $p - 1$ other processes all by itself, the time complexity of the process creation is $\Theta(p)$. Assuming that there are p processors and p memory banks and each process runs on its own processor, then there are no memory conflicts. Hence the time taken to come up with the local sums is $\Theta(n/p)$. Since *global_sum* is updated inside a critical section, only one process can have access to it at any particular time. Hence the processes must "line up" for the resource, and the complexity of the updating is $\Theta(p)$. The final endfor is where the synchronization occurs. If synchronization is done via a single global semaphore, then again each processor must have exclusive access to the semaphore, and the complexity of synchronization is $\Theta(p)$. Hence the complexity of this algorithm is $\Theta(n/p + p)$. The minimum of the function $n/p + p$ occurs when $p = \sqrt{n}$. Hence the execution time of the parallel algorithm is minimized when $\Theta(\sqrt{n})$ processors are used.

3-4 PROCESS COMMUNICATION AND SYNCHRONIZATION ON MIMD MODELS

This section addresses a second fundamental question related to solving problems on MIMD computers: What mechanisms allow processes to work together, and how are these mechanisms expressed in a programming language? Fortunately, much is already known about managing concurrent processes, since these issues are addressed in multiprogrammed operating systems: how to express concurrency, how to allow communication between processes, and how to synchronize processes. The primary source of material for this section is the survey article of Andrews and Schneider [1983].

Communication and Synchronization

In order for processes to work together, they must have the ability to communicate and synchronize. Communication is achieved either through shared variables or through message passing. The type of communication available may be dictated by architecture. For example, multicomputers do not have any shared memory; hence they must achieve communication through message passing.

Communication often leads to synchronization requirements. Synchronization has two uses: to constrain the ordering of events and to control interference. For example, consider a pipelined parallel algorithm in which one process is writing into a buffer and another process is reading from it. Clearly a process cannot write into a full buffer. Likewise, a process cannot read from an empty buffer. A mechanism to prevent these events from happening is an example of the first kind of synchronization.

The second use of synchronization is to control interference. Recall the partitioned algorithm described above that finds the sum of n values. In order for the result to be correct, each process must be able to add its local sum to the global sum as an atomic operation. The use of the lock is to prevent a process from accessing the value of the global sum while another process is updating it.

Expressing Concurrency

A number of programming language constructs to express concurrency have been proposed. Although these constructs were developed to express concurrency in multiprogrammed operating systems, they can be used in a parallel programming environment.

Fork and Join. The fork statement is similar to a procedure call in the sense that it enables the commencement of a particular routine. However, unlike a procedure call, the calling process continues execution. Hence a single process is "forked" into two processes. The invoking process can synchronize with termination of the forked process by executing the join statement. When the invoking process encounters the join statement, it cannot continue execution until the invoked process has terminated.

For example, consider the following program:

```
program a;          program b;
    ...                 ...
    fork b;             ...
    ...                 ...
    join b;             ...
    ...                 end;
```

A single process executes program a until the fork statement is encountered. At this point, execution of a continues and execution of program b begins. If the process executing a reaches join b before the process executing program b terminates, then the first process suspends execution. After the forked process terminates execution of program b, the first process may continue by executing the statements after the join.

The powerful fork and join constructs have been widely used. For example, the UNIX[†] operating system makes frequent use of variants of fork and join [Ritchie and Thompson 1974]. However, since fork and join may appear anywhere in a program (e.g., in loops or in conditional statements), programs making unstructured use of these constructs may be difficult to understand and debug.

Cobegin and Coend. In contrast to the unstructured fork and join statements, cobegin and coend are a structured way of indicating a set of statements that can be executed in parallel. The statement

$$\text{cobegin } S_1 || S_2 || S_3 || \ldots || S_n \text{ coend}$$

causes statements $S_1, S_2, S_3, \ldots, S_n$ to be executed concurrently. Each of these statements may be a block or another cobegin. Processing continues beyond the coend only after all the S_i's have completed execution.

The cobegin/coend construct is not as powerful as fork/join. Any concurrent program can be represented by a **process flow graph**, an acyclic, directed graph in which vertices represent processes and edges represent execution constraints (i.e., an edge from i to j indicates that process j cannot begin execution until process i has finished execution) [Shaw 1974]. "Without introducing extra processes or idle time, cobegin and sequencing can only represent series-parallel (properly nested) process flow graphs. Using fork and join, the computation represented by any process flow graph can be specified directly" [Andrews and Schneider 1983]. A second limitation of cobegin is that, at least as it has been

---

[†] *UNIX* is a trademark of Bell Laboratories.

implemented in current languages, only a fixed number of processes may be enabled.

Dijkstra [1968b] invented this construct; he first called it **parbegin**. Variants of **cobegin/coend** have appeared in the programming languages ALGOL68 [van Wijngaarden et al. 1975], Communicating Sequential Processes [Hoare 1978], Edison [Brinch Hansen 1981], and Argus [Liskov and Scheifler 1982].

Process Declarations. It is common for a concurrent program to be made up of a number of sequential procedures operating concurrently. Such a program can be written by using **fork** or **cobegin**, but the use of process declarations makes the structure of the program easier to understand by allowing the programmer to specify in a procedure's declaration that it should be executed concurrently.

Some programming languages allow exactly one instance of each declared process. Hence a set of declared processes is equivalent to a single **cobegin**, where each statement in the **cobegin** is one of the declared processes. Distributed Processes [Brinch Hansen 1978] and SR [Andrews 1981] declare processes in this way.

Other programming languages allow more than one instance of a declared process to execute concurrently. Some of these, such as Concurrent Pascal [Brinch Hansen 1975] and Modula [Wirth 1977a], require that all processes be activated at the beginning of the program's execution. Others, such as PLITS [Feldman 1979] and Ada [U.S. Department of Defense 1981], allow processes to be created at any time during the execution of the concurrent program, which means that the number of processes may vary from execution to execution.

Synchronizing with Shared Variables

Mutual exclusion and condition synchronization are two frequently used types of synchronization on systems in which processes communicate via shared variables. A **critical section** is a sequence of statements that must appear to be executed as an atomic operation. Recall the parallel sum-finding algorithm presented earlier. Each process finds a partial sum and then executes the statement

$$global\_sum \leftarrow global\_sum + local\_sum$$

If no mechanism were used to ensure the atomicity of this statement, then the correctness of the final result would be in jeopardy. If process A read the current value of *global_sum* and performed the addition, resulting in a new value, but did not store this result before process B read the value of *global_sum*, then the outcome of the entire computation would be erroneous. Hence this statement is an example of a critical section. **Mutual exclusion** refers to the mutually exclusive execution of critical sections.

Condition synchronization is a method for delaying the continued execution of a process until some data object it shares with another process is in an appropriate state. For example, consider a pipelined algorithm for implementing a compiler. Imagine that the first process tokenizes the input and passes these tokens to the second process, the parser. Suppose that a buffer is used to hold the tokens generated by the tokenizer, but not yet consumed by the parser. Obviously the tokenizer should be delayed when the buffer is full. Similarly the parser must be delayed if the buffer is empty.

There are a number of different ways to implement synchronization on a system that uses shared variables for communication. The following paragraphs describe three methods: busy-waiting, semaphore, and monitors.

Busy-waiting. A simple way to implement synchronization is to have processes test and set shared variables. To signal a condition, a process sets the value of a shared variable. To wait for a condition, a process repeatedly tests the value of a shared variable. Because a waiting process must continue to use processor cycles to test the value of the variable, this technique is called **busy-waiting**. A **spin lock** is a shared variable used this way; a process waiting for the value of a spin lock to change is said to be **spinning**.

Two solutions to the two-processor mutual-exclusion problem using busy-waiting are Dekker's algorithm [Shaw 1974] and a simpler algorithm by Peterson [1981].

Busy-waiting has a number of disadvantages. Spinning processes waste processor cycles. Busy-waiting protocols are hard to understand, debug, and prove correct, in part because they are implemented at such a low level.

Semaphores. A **semaphore** is a variable defined over the nonnegative integers with two associated operations, P and V. Given a semaphore s, a process executing P(s) is delayed until $s > 0$, at which time it executes $s \leftarrow s - 1$ and continues. The test and decrement must be performed as an indivisible operation. A process executing V(s) performs $s \leftarrow s + 1$ as an indivisible operation and then continues.

Semaphores implement synchronization at a higher level than busy-waiting, and they have been widely used since their formal introduction by Dijkstra [1968a, 1968b]. Operation P comes from the Dutch word *passeren*, meaning "to pass"; V comes from the Dutch word *vrygeven*, meaning "to release." Some authors use wait and signal instead of P and V.

Semaphores can be used to implement a wide variety of synchronization functions. For example, to implement mutual exclusion, every critical section is preceded by a P operation and followed by a V operation on the same semaphore. Critical sections that must be mutually excluded use the same semaphore, which begins with the value 1. In such an implementation the semaphore always has the value 0 or 1; hence it is called a **binary semaphore**.

In order to implement condition synchronization by using a semaphore, the semaphore begins with the value 0, assuming that the condition is initially false. After a process has made the condition true, it signals this by executing the V operation on a semaphore. A process waiting for the condition to be true executes the P operation on the same semaphore.

Although semaphores can be implemented by busy-waiting, they are usually implemented through system calls to the **kernel**, the part of the operating system that schedules processes. The kernel of a multiprogrammed operating system keeps a number of lists containing process descriptors. Ready-to-run processes are kept on the **ready list**. Processes that are blocked, waiting to perform a P operation on a semaphore, are kept on a list associated with that semaphore. Since the kernel gives time slices only to processes whose descriptors are on the ready list, the processor never wastes time trying to execute blocked processes. This is an advantage over the busy-waiting approach. When a process encounters a P or V operation, execution traps to a routine in the kernel. This guarantees (1) indivisibility and (2) access to the lists of process descriptors. In the case of P, if the semaphore is positive, it is decremented and the process continues execution. If the semaphore is zero, the process is added to the queue of processes blocked on that semaphore. In the case of V, if the queue of processes blocked on that semaphore is nonempty, the process descriptor at the head of the queue is removed and added to the ready list. If the blocked queue is empty, the semaphore is incremented.

Implementing a kernel on an MIMD computer is more complicated. One approach is for a single processor to maintain the ready list and allocate processes to the other processors. This idea is clearly unacceptable in the case of multicomputers. Another approach is for multiple processors to share a single ready list. Concurrent access to this list is a problem that must be addressed. Mutual exclusion to the list can be implemented by using busy-waiting. Busy-waiting is appropriate, because the critical section is short and because a processor accesses the ready list since it is looking for something to do. Therefore, it has nothing better to do than busy-wait.

Sharing a single ready list implies a shared memory; hence this method, too, cannot be implemented on a multicomputer. On a multicomputer it is more appropriate for each processor to have its own kernel and its own ready list. Processes migrating from processor A to processor B would leave the jurisdiction of processor A's kernel and enter the jurisdiction of processor B's kernel.

Monitors. A disadvantage of semaphores is that they are unstructured operations. This lack of structure can lead to errors. For example, every critical section must begin with a P operation and end with a V operation on the same semaphore. If one of these operations is forgotten, or if the two operations access different semaphores, or if two critical regions that are supposed to be mutually exclusive do not reference the same semaphore, then mutual exclusion is no longer guaranteed. The monitor is a structured way of implementing mutual exclusion.

A **monitor** consists of variables representing the state of some resource, procedures that implement operations on that resource, and initialization code. The initialization code initializes the values of variables before any procedure in the monitor is called; these values are retained between procedure invocations and may be accessed only by procedures in the monitor. Monitor procedures resemble ordinary procedures in the programming languages with one significant exception: The execution of procedures in the same module is guaranteed to be mutually exclusive. Hence monitors are a structured way of implementing mutual exclusion. Condition synchronization by using monitors is accomplished through the use of semaphores by the operating system.

Monitors are implemented by using mutual exclusion. Programming languages that support monitors include Concurrent Pascal [Brinch Hansen 1975, 1977] and Modula [Wirth 1977a, 1977b, 1977c].

Low-Level Synchronization through Message Passing

One way processes can communicate and synchronize is through the use of shared variables. Another way is through message passing. Message passing is a form of communication, since a process receiving a message is receiving values from another process. Message passing is a form of synchronization, since a message can be received only after it has been sent.

In order to send a message, a process executes a statement of the form

send expression_list to destination_designator

The values of the expressions in the expression list are put into a message that is sent to the designated destination. In order to receive a message, a process executes a statement of the form

receive variable_list from source_designator

When a process receives a message from the designated source, it assigns the values in the message to the variables in the variable list and then destroys the message.

One way in which message passing schemes differ is the means for specifying the source and destination designators. The simplest method is for the source and destination designators to be the names of processes. This is called **direct naming**, and it is easy to implement and use. Direct naming is a good way to implement pipelined algorithms, since the flow of information is constant throughout program execution.

Another way to implement message passing is for source and destination designators to refer to **global names**, or **mailboxes**. In the general case, many processes can send messages to a mailbox, and many processes can receive messages from that mailbox. This allows the construction of client-server programs, in which a number of client processes make requests that are handled by a number of server processes. A simpler variant on mailboxes, called **ports**, allows multiple client processes but only a single server process per port [Balzer 1971].

A second way in which message passing schemes differ is when the source and destination designators are decided. If source and destination designators are fixed at compile time (**static channel naming**), then programs may use only channels that are known at compile time, which limits their flexibility in a changing environment. Static channel naming also implies that a process must have permanent access to a channel, even if it requires use of it for only a short period. It is often better to allocate channels as a resource.

Dynamic channel naming delays the computation of source and destination designators until run time. Dynamic channel naming can be implemented with a static channel-naming scheme that uses variables to contain source or destination designators.

A third attribute of systems based on message passing is whether sending or receiving a message can delay the further execution of the process performing the statement. A **nonblocking** statement never delays the further execution of the invoking process. Otherwise, the statement is said to be **blocking**. If there is no buffer between a process sending a message and a process receiving a message, then both send and receive are blocking: the send delays until a corresponding receive is executed. At that time the message is passed, and then both the sending and the receiving processes continue execution. This is called **synchronous message passing**.

If buffers are used to store messages that have been sent but not yet received (**buffered message passing**), then the system designer has a number of options. If a send is executed when the buffer is full, the send might delay until there is room in the buffer. Alternatively, a condition code could be returned to the sender that the message could not be sent, owing to the full buffer. Similarly, if a receive is executed when the buffer is empty, then either it delays until there is a message in the buffer or it returns a condition code that the buffer was empty.

Finally, if the buffer between a sending process and a receiving process is effectively unbounded, then send is nonblocking. This protocol is called either **asynchronous message passing** or **send no-wait**. Asynchronous message passing allows the sending process to get arbitrarily far ahead of the receiving process.

Remote Procedure Call

When a client-server interaction is programmed, each of the two processes sends a pair of messages. The client performs a send immediately followed by a receive; the server executes a receive, performs the service, and then executes a send. The **remote procedure call** is a higher-level construct that supports this kind of interaction. A remote procedure call allows a client process to call upon a server process, which may or may not be on the same processor. The syntax of a remote procedure call is similar to that of an ordinary procedure call; the process requesting service executes a statement of the form

call *service*(*value_arguments*; *result_arguments*)

The *service* is the name of a channel. If direct naming is used, *service* refers to the server process; otherwise, it refers to the kind of service desired.

When a remote procedure call is executed, the value arguments are sent to the server process. The client process is delayed until the values of the result arguments have been returned by the server.

Two approaches have been taken to specifying the server process of a remote procedure call. The first approach is to declare a server similarly to ordinary procedures in a sequential programming language:

```
remote procedure service
    (in value_parameters; out result_parameters)
    body
end
```

Although declared as a procedure, it is implemented as a process. The process idles until it receives a message containing the value parameters from a calling process. The body of the procedure is executed, and then the values of the result parameters are passed back to the calling process via a **reply message**. Either this kind of server can be implemented as a single looping process that handles one remote procedure call at a time, or a new server process can be created for each remote procedure call.

The second approach taken to specifying the server process is to treat the remote procedure as a statement. The statement has the form

```
accept service(in value_parameters; out result_parameters) → body
```

A server process executing this statement delays until it receives a message from a remote procedure call requesting the service. At this time the value parameters become available, and the body of the statement is executed. After execution of the statement body has completed, a reply message, which contains the values of the result parameters, is sent to the calling process, and the server continues execution with the statement following the accept. "When accept or similar statements are used to specify the server side, remote procedure call is called a **rendezvous** [U.S. Department of Defense 1981], because the client and server 'meet' for the body of the accept statement and then go their separate ways" [Andrews and Schneider 1983].

Using rendezvous has two advantages. First, it allows server processes to choose when they will service remote procedure calls. Second, by using more than one accept statement, one server can provide different kinds of service.

Classifying Concurrent Programming Languages

Programming languages allowing concurrency can be divided into three categories: procedure-oriented languages, message-oriented languages, and operation-oriented languages. It is conceivable that any category of concurrent programming language could be implemented on any kind of MIMD computer, but implementing a language that does not match the architecture will do little to help the programmer develop efficient parallel algorithms.

"In **procedure-oriented languages**, process interaction is based on shared variables" [Anderson and Schneider 1983]. Processes have direct access to the data they want to manipulate. Hence concurrent access to shared data is possible. Therefore, the programming language must provide some way of ensuring mutual exclusion of processes in critical sections. Concurrent Pascal, Modula, Mesa, and Edison are examples of procedure-oriented languages. Since processes manipulate data directly, this kind of language is most appropriate for multiprocessors, either tightly coupled or loosely coupled.

Message-oriented languages are built upon the primitives **send** and **receive**. Unlike procedure-oriented languages, message-oriented languages do not give processes access to every data object. Every object has a caretaker process, which manages it. In order to manipulate an object, a process must send a message to its caretaker. Hence concurrent access is not an issue in message-oriented languages. Communicating Sequential Processes, Gypsy, occam, and PLITS are examples of message-oriented languages. When it is implemented on a multicomputer, a message-oriented language can make the communication network transparent, simplifying the programmer's job.

Operation-oriented languages use the remote procedure call as the principal means for interprocess communication and synchronization. Operation-oriented languages combine aspects of the other two categories. Operations are performed on objects by calling procedures, as in procedure-oriented languages. However, objects are managed by caretakers, as in message-oriented languages. Examples of operation-oriented languages include Distributed Processes, StarMod, Ada, and SR. Operation-oriented languages can be implemented efficiently on multiprocessors and multicomputers.

3-5 DEADLOCK

A set of active concurrent processes is said to be **deadlocked** if each holds non-preemptible resources that must be acquired by some other process in order for it to proceed. The potential for deadlock exists whenever multiple processes share resources in an unsupervised way. Hence deadlock can exist in multiprogrammed operating systems as well as in multiprocessors and multicomputers, and designers of parallel algorithms can benefit from this experience.

As an illustration of deadlock, consider the two processes executing in Figure 3-7. Each process attempts to perform a lock operation on two resources. Note that lock and unlock correspond to P and V operations on binary semaphores. Process 1 locks A while process 2 locks B. Process 1 is blocked when it tries to lock B; likewise, process 2 is blocked when it tries to lock A. Neither process can proceed. If one of the processes cannot be made to "back up" and yield its semaphore, the two processes will remain deadlocked indefinitely.

Four conditions are necessary for a deadlock to exist [Coffman and Denning 1973]: (1) mutual exclusion: each process has exclusive use of its resources;

Process 1	Process 2
.	.
.	.
.	.
lock(A);	lock(B)
.	.
.	.
lock(B);	lock(A)

Figure 3-7 Illustration of deadlock. As long as each process has exclusive access to a resource desired by the other, neither process can proceed.

(2) nonpreemption: a process never releases resources it holds until it is through using them; (3) resource waiting: each process holds resources while waiting for other processes to release theirs; (4) a cycle of waiting processes: each process in the cycle waits for resources that the next process owns and will not relinquish.

The problem of deadlock is commonly addressed in one of three ways. One approach is to detect deadlocks when they occur and try to recover from them. Another approach is to avoid deadlocks by using advance information about requests for resources to control allocation, so that the next allocation of a resource will not cause processes to enter a situation in which deadlock may occur.

The third approach is to prevent deadlock by forbidding one of the first three conditions listed above. This can be accomplished in a number of ways. A cycle of waiting processes can be prevented by ordering shared resources and forcing processes to request resources in that order. Deadlock could have been prevented in the example of Figure 3-7 if each process had attempted to lock A before locking B. Deadlock can also be prevented by requiring processes to acquire all their resources at once. The second approach often leads to underutilization of resources, however, since it may require processes to retain possession of resources long before and after the resources are being used.

3-6 TASK SCHEDULING ON MIMD COMPUTERS

The primary goal of the parallel algorithm designer is to come up with an algorithm that executes as quickly as possible. This section addresses the crucial question of task scheduling: How should tasks be allocated to processors so that the execution time of the algorithm is minimized?

Deterministic Models

A **schedule** is an allocation of tasks to processors. In a **deterministic model**, the precedence relations between the tasks and the execution time needed by

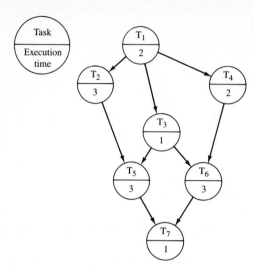

Figure 3-8 A task graph. Each node represents a task to be performed. A directed arc from T_i to T_j indicates that task T_i must complete before task T_j begins.

each task are fixed and known before the schedule is devised. Deterministic models are unrealistic. They ignore variances in the execution times of tasks due to interrupts, contention for shared memory, etc. But at least they make possible the static allocation of tasks to processors.

An optimal schedule minimizes the total execution time of the parallel algorithm and the number of processors used. Even quite simple instances of the scheduling problem are often intractable. For example, consider the task graph illustrated in Figure 3-8. We are given a set of seven tasks and precedence relations among the tasks (i.e., information about what tasks must be completed before other tasks can be started). Furthermore, the execution time needed by each task is fixed and known in advance. Three processors are available. Scheduling the tasks on the processors to minimize the time needed to complete all the tasks is an NP-hard problem, meaning it is unlikely that a polynomial-time algorithm exists that can always find an optimal schedule, given an arbitrary task graph. Therefore, in general we must be content with scheduling algorithms that do a good, but not perfect, job in polynomial time.

Theorem 3-1 (See Graham [1972].) Given a set of p identical processors and a deterministic model, any scheduling of processes to processors results in a schedule whose time is not more than twice that required by the optimal schedule.

Deterministic schedules are often illustrated with **Gantt charts**. A Gantt chart indicates the time each tasks spends in execution as well as the processor on which it executes. For example, Figure 3-9 is a Gantt chart of a schedule based on the precedence graph of Figure 3-8. A desirable feature of Gantt charts is that they graphically illustrate the **utilization** of each processor (percentage of time spent executing tasks).

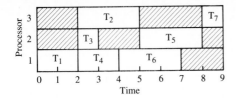

Figure 3-9 A Gantt chart illustrating a schedule for the task graph of Figure 3-8.

Nondeterministic Models

In a **nondeterministic model**, the execution time of a task is represented by a random variable, making the scheduling problem more difficult. This subsection summarizes mathematics developed by Robinson [1979] that allows the estimation of the expected execution time of parallel programs with "simple" task graphs. Note that Robinson's work assumes the tightly coupled multiprocessor model.

Let us formalize the notion of a task graph. Given a set of tasks $T_1, T_2, \ldots,$ T_n and a partial ordering $<$, task T_i is a **predecessor** of task T_j and T_j is a **successor** of T_i if $T_i < T_j$. Tasks with no predecessors are called *initial tasks*.

A set of tasks is *independent* if for every pair of tasks T_i and T_j in the set, neither is a predecessor of the other. The *width* of a task graph is the size of the maximal set of independent tasks.

A *chain* is a totally ordered task graph. The *length* of a chain is the number of tasks in the chain. The *level* of a task T in a task graph G is the maximum chain length in G from an initial task to T. The *depth* of a task graph G is the maximum level of any task in G.

Definition 3-1 (See Robinson [1979].) Given a task graph G, let $C_1, C_2,$ $\ldots, C_m$ be all the chains from initial to final tasks in G. For every chain C_i consisting of tasks $T_{i_1}, T_{i_2}, \ldots, T_{i_j}$, let X_i be the expression $x_{i_1}, x_{i_2}, \ldots, x_{i_j}$, where $x_1, x_2, \ldots, x_n$ are polynomial variables. Then G is a *simple task graph* if the polynomial $X_1 + X_2 + \cdots + X_m$ can be factored so that every variable appears exactly once.

Figure 3-10 illustrates simple and nonsimple task graphs. Robinson has observed that the set of simple task graphs corresponds exactly to the set of task graphs that can be generated by parallel languages whose only concurrent programming construct is a **cobegin/coend** statement, or its equivalent.

Theorem 3-2 (See Robinson [1979].) Given a simple task graph G, if the number of processors exceeds the width of G, the tasks are independent, the depth of G is L, and the execution time of all m_j tasks on level j is a random variable with mean μ_j and standard deviation σ_j, then

$$\sum_{1 \leq j \leq L} \mu_j \leq E(t_G) \leq \sum_{1 \leq j \leq L} \left(\mu_j + \frac{m_j - 1}{\sqrt{2m_j - 1}} \sigma_j \right)$$

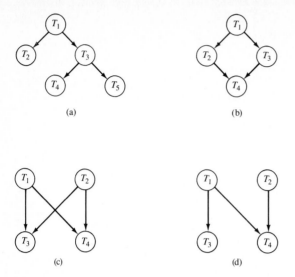

Figure 3-10 Simple and nonsimple task graphs. (a) Simple task graph: $x_1x_2 + x_1x_3x_4 + x_1x_3x_5 = x_1[x_2 + x_3(x_4 + x_5)]$. (b) Simple task graph: $x_1x_2x_4 + x_1x_3x_4 = x_1(x_2 + x_3)x_4$. (c) Simple task graph: $x_1x_3 + x_1x_4 + x_2x_3 + x_2x_4 = (x_1 + x_2)(x_3 + x_4)$. (d) Nonsimple task graph.

For example, consider the following hypothetical parallel algorithm to sort $n = pk$ values by using p processors. The partitioned algorithm has two phases. In the first phase each of p processors sorts $k = n/p$ values. After all the processors synchronize, each processor merges $k = n/p$ values, completing the sort. Suppose that the sort used in the first phase is the best known sequential sort, with an expected execution time of $3n\log n + 200$ and a standard deviation of $2\sqrt{n}$ to sort n values. Furthermore, suppose that the merge algorithm merges n values with an expected execution time of $4n\log n + 100$ and a standard deviation of $\sqrt{n}$. According to the theorem,

$$E(t_G) \geq 3\frac{n}{p}\log\frac{n}{p} + 200 + 4\frac{n}{p}\log\frac{n}{p} + 100$$

$$= 7\frac{n}{p}\log\frac{n}{p} + 300$$

$$E(t_G) \leq 7\frac{n}{p}\log\frac{n}{p} + 300 + \frac{(p-1)3\sqrt{n}}{\sqrt{2p-1}}$$

The expected speedup can be computed by dividing the expected execution time of the sequential algorithm by the expected execution time of the parallel algorithm. Since the sorting algorithm used in the first phase is the fastest known sequential sort, the expected execution time of the sequential algorithm is $3n\log n + 200$. The expected execution time of the parallel algorithm is not

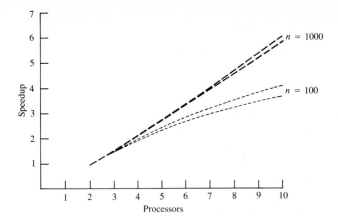

Figure 3-11 Expected speedup of hypothetical parallel sort algorithm. Note that for each problem size the predicted speedup is a range between two lines.

known exactly; it falls within a range. Figure 3-11 graphs the expected speedup of the parallel algorithm for two list sizes, 100 and 1000.

Note that Robinson's result does not take into account factors that increase the execution time of a parallel algorithm, such as process creation time, contention for shared variables, and contention for the ready list.

3-7 SUMMARY

This chapter has introduced some factors to be considered in designing parallel algorithms. The character of the algorithm depends heavily on the architecture on which it is to be implemented.

The designer of algorithms for processor arrays must be aware that the communication complexity of a parallel algorithm can easily dominate the computational complexity.

MIMD computers, being composed of asynchronous processors executing different instruction streams, give the algorithm developer more options. Algorithms for MIMD computers can be put into one of three general categories—pipelined, partitioned, or relaxed—or they may have characteristics of all three categories. Many factors must be taken into consideration in designing an algorithm for an MIMD computer, if the processors are to be efficiently used. The factors include the number of process creations and synchronizations, amount of sequential code, contention for shared data structures, distribution of the workload, and efficiency of the underlying parallel algorithm.

Processes work together by communicating and synchronizing with each other. Communication can be achieved either through shared variables or through message passing. A number of synchronization mechanisms have been proposed for shared-variable systems, including busy-waiting, semaphores, and monitors. Message passing achieves both communication and synchronization. The remote procedure call is a structured way of implementing client-server programs. Programming languages allowing concurrency can be divided into three categories: procedure-oriented languages, message-oriented languages, and operation-oriented languages.

Deadlock can occur whenever multiple processes share resources in an unsupervised fashion. The problem of deadlock is addressed by attempting to detect and recover from deadlocks when they occur, by avoiding situations that may lead to deadlock, or by preventing deadlock by making one of the conditions necessary for deadlock impossible.

A schedule is an allocation of parallel processes to physical processors. The goal of a scheduler is to minimize the time to completion of a parallel algorithm. Deterministic scheduling models assume that the computation time for each task and the precedence relations between pairs of tasks are fixed and known when the schedule is devised. The execution time of a task in a nondeterministic model is a random variable, making scheduling more difficult.

BIBLIOGRAPHIC NOTES

Agerwala and Lint [1978], Lint and Agerwala [1981], and Irani and Chen [1982] have discussed the importance of communications issues in the design of parallel algorithms. Nassimi and Sahni [1980a] present an algorithm for routing data on SIMD-MC computers. Their algorithm is optimal for any data routing that can be specified in terms of permutating and complementing the address bits of a processing element. Examples of such data routings include matrix transpose, bit reversal, vector reversal, and perfect shuffle. Raghavendra and Prasanna Kumar [1986] present optimal algorithms for performing a wide variety of permutations on a two-dimensional mesh with wraparound connections.

The design and analysis of algorithms for MIMD computers are discussed by Baudet [1978a, 1978b], Jones and Gehringer [1980], Kung [1976, 1980], Oleinick [1982], Raskin [1978], and Robinson [1977]. Our categorization of algorithms for MIMD computers is not new. Earlier works using similar categorizations include Jones and Schwarz [1980] and Gilbert [1983]. Satyanarayanan [1980] has written an annotated bibliography on multiprocessing. Rodeheffer and Hibbard [1980] describe attempts to automatically detect parallelism in ordinary programs and exploit it on MIMD computers. Kasahara and Narita [1984] discuss scheduling algorithms for MIMD computers.

Joseph et al. [1984] have devoted a textbook to the topic of implementing a multiprocessor operating system.

A good source of articles and further references on parallel programming languages is the August 1986 issue of *Computer*.

EXERCISES

3-1 Argue against renaming this chapter "Making Algorithms Parallel."

3-2 Is integer arithmetic on a digital computer associative? Is it commutative? Is floating-point arithmetic associative? Is it commutative? Justify your answer.

3-3 Reconsider the sum-finding algorithm for the SIMD-PS model. What processing elements contain the sum when the algorithm terminates?

3-4 Write an $\Theta(p)$ parallel algorithm to perform the equivalent of a shuffle operation on the SIMD-MC^1 model. In other words, simulate a shuffle operation on a one-dimensional mesh of $p = 2^k$ processors.

3-5 What is a lower bound on the complexity of a parallel algorithm to find the sum of $n = l^3$ integers on the SIMD-MC^3 model, assuming that initial values are distributed evenly among the processing elements?

3-6 Answer the previous question, changing the model to SIMD-MC^2.

3-7 Write an algorithm to add $n^{3/2}$ values in $\Theta(\sqrt{n})$ time on an SIMD-MC^2 model with n processing elements organized in a $\sqrt{n} \times \sqrt{n}$ mesh. Assume each processing element initially contains $\sqrt{n}$ values.

3-8 Design an $\Theta(\sqrt{n})$ algorithm for finding the sum of n integers on a $\sqrt{n} \times \sqrt{n}$ mesh with the property that at the end of the computation, every processing element contains the sum.

3-9 Monitors have been proposed as a structured alternative to semaphores. However, since monitors are implemented by using semaphores, in what way do monitors represent an improvement in the programming environment?

3-10 Is a multiprogrammed operating system running on a multiprocessor an implementation of a parallel algorithm?

3-11 Devise a Gantt chart showing a two-processor schedule for the task graph of Figure 3-8 that requires 9 units of time to execute.

3-12 Given a task graph and an arbitrarily large number of processors, what is a lower bound on the length of an optimal schedule?

3-13 Map nodes of a complete binary tree with four levels (Figure 3-12a) to nodes of a four-dimensional hypercube (Figure 3-12b) so that as many tree edges as possible map onto hypercube edges.

*3-14 Prove that a complete binary tree with n levels ($2^n - 1$ nodes) can be embedded in an $(n + 1)$-dimensional hypercube. [In other words, prove that an $(n + 1)$-dimensional hypercube has a complete binary tree with n levels as a subgraph.]

3-15 Embed a 4×4 mesh with wraparound connections into a four-dimensional hypercube.

*3-16 Given $n = 2^{k/2}$ for a positive even integer k, write a function that maps elements of an $n \times n$ mesh with wraparound connections into nodes of a

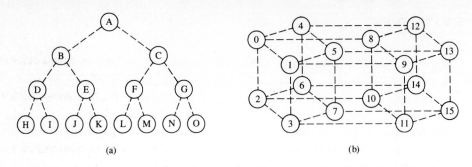

Figure 3-12 Figure for Exercise 3-13. (a) Complete binary tree with four levels. (b) Four-dimensional hypercube.

k-dimensional hypercube so that as many mesh connections as possible map directly onto connections in the hypercube.

*3-17 Prove that the smallest hypercube containing an embedded $l_1 \times l_2 \times \cdots \times l_k$ mesh has dimension $d_1 + d_2 + \cdots + d_k$, where $d_i = \lceil \log l_i \rceil$ for $1 \leq i \leq k$. Assume that the mesh does not contain wraparound connections.

FOUR

SORTING AND THE FAST FOURIER TRANSFORM

Sorting is one of the most common activities performed on serial computers. Many algorithms incorporate a sort so that later accesses of information may be done efficiently. Sorting has additional importance to designers of parallel algorithms: it is frequently used to perform general data permutations, including random access read and random access write. These data movement operations can be used to solve problems in graph theory, computational geometry, and image processing in optimal or near optimal time.

Much work has been done developing parallel sorting algorithms. This chapter describes a variety of sorting algorithms for processor arrays and multiprocessors. All these algorithms are **internal sorts**—that is, they sort tables small enough to fit entirely in primary memory. In addition, all the algorithms described below sort by comparing pairs of elements.

Section 4-1 presents Batcher's bitonic merge algorithm, which has been the basis for many other parallel sorting algorithms. Section 4-2 describes sorting algorithms for processor arrays, including the odd-even transposition sort for linear arrays and bitonic merge sort for mesh, cube-connected, and perfect shuffle networks. The implementation of a parallel quicksort algorithm on multiprocessors is the subject of Section 4-3.

The fast Fourier transform has much in common with sorting algorithms. Section 4-4 describes the fast Fourier transform and an implementation of the algorithm on a cube-connected processor array.

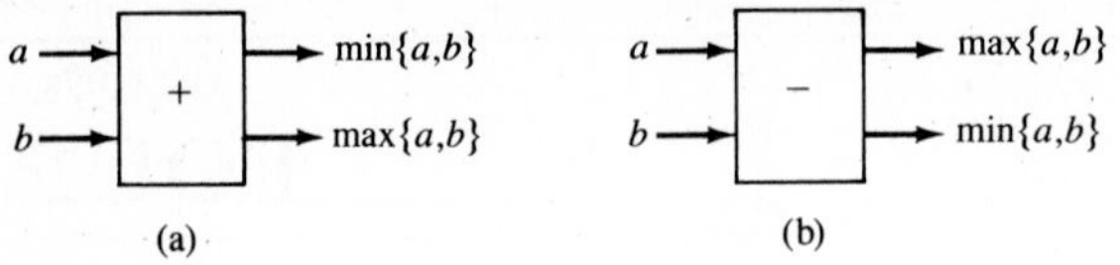

Figure 4-1 Two comparators.
(a) Low-to-high comparator.
(b) High-to-low comparator.

4-1 BITONIC MERGE

Assume that we are given a table of n elements, denoted $a_0, a_1, \ldots, a_{n-1}$, on which a linear order has been defined. Thus for any two elements a_i and a_j, exactly one of the following cases must be true: $a_i < a_j$, $a_i = a_j$, or $a_i > a_j$. The goal of sorting is to find a permutation $(\pi_1, \pi_2, \ldots, \pi_n)$ such that $a_{\pi_1} \leq a_{\pi_2} \leq \cdots \leq a_{\pi_n}$. It is well known that any sequential sorting algorithm based upon comparisons of pairs of elements must have complexity $\Omega(n \log n)$.

In 1968 Batcher introduced a parallel sorting algorithm with time complexity $\Theta(\log^2 n)$ [Batcher 1968]. This algorithm, called **bitonic merge**, has been the basis for sorting algorithms on several models of parallel computation. The fundamental operation is called **compare-exchange**: two numbers are routed into a **comparator**, where they are exchanged, if necessary, so that they are in the proper order (Figure 4-1).

Definition 4-1 A **bitonic sequence** is a sequence of numbers $a_0, \ldots, a_{n-1}$ with the property that (1) there exists an index i, $0 \leq i \leq n-1$, such that a_0 through a_i is monotonically increasing and a_i through a_{n-1} is monotonically decreasing, or else (2) there exists a cyclic shift of indices so that the first condition is satisfied.

A single compare-exchange step can split a single bitonic sequence into two bitonic sequences. Assume $n = 2^k$ for some integer $k > 0$. Let $(a_0, a_1, \ldots, a_{n-1})$ be a bitonic sequence with the property that $a_0 \leq a_1 \leq \cdots \leq a_{n/2-1}$ and $a_{n/2} \geq a_{n/2+1} \geq \cdots \geq a_{n-1}$. (Actually, the length of the ascending portion of the sequence need not be equal to the length of the descending portion of the sequence for the sequence-splitting compare-exchange step to work correctly. However, this assumption makes the explanation easier to follow.) Now consider the two sequences

$$\min(a_0, a_{n/2}), \ \min(a_1, a_{n/2+1}), \ \ldots, \ \min(a_{n/2-1}, a_{n-1})$$

and

$$\max(a_0, a_{n/2}), \ \max(a_1, a_{n/2+1}), \ \ldots, \ \max(a_{n/2-1}, a_{n-1})$$

Both sequences are bitonic. In addition, every element in the first sequence is smaller than every element in the second sequence. Hence a single compare-exchange step transforms a single bitonic sequence into two bitonic sequences half as long as the original sequence (see Figure 4-2). It is easy to see, then, that a bitonic sequence of length 2^k can be sorted in k steps by recursively applying the same compare-exchange operation to the resulting bitonic sequences. Figure 4-3

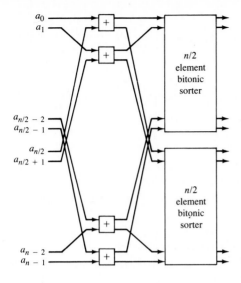

Figure 4-2 The recursive nature of bitonic merge. Given a bitonic sequence, a single compare-exchange step divides the sequence into two bitonic sequences of half the length. Applying this step recursively yields a sorted sequence, which can be thought of as half of a bitonic sequence of twice the length.

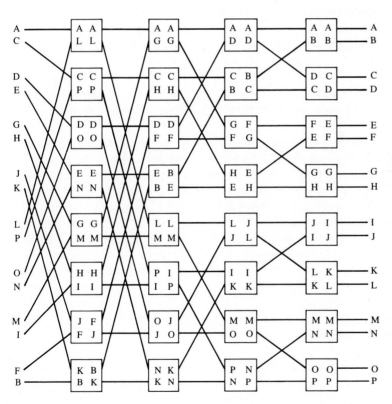

Figure 4-3 Sorting a bitonic sequence of length 16 by using bitonic merge.

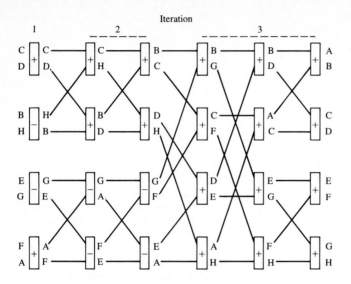

Figure 4-4 Bitonic mergesort of an unsorted list of eight elements.

illustrates how a bitonic sequence of length 16 is sorted in four compare-exchange steps.

Theorem 4-1 (See Batcher [1968].) A list of $n = 2^k$ unsorted elements can be sorted by using a network of $2^{k-2}k(k+1)$ comparators in time $\Theta(\log^2 n)$.

Proof. Bitonic merge takes a bitonic sequence and transforms it into a sorted list, *which can be thought of as half a bitonic sequence of twice the length.* If a bitonic sequence of length 2^m is sorted into ascending order, while an adjacent sequence of length 2^m is sorted into descending order, then after m compare-exchange steps the combined sequence of length 2^{m+1} is a bitonic sequence. A list of n elements to be sorted can be viewed as a set of n unsorted sequences of length 1 or as $n/2$ bitonic sequences of length 2. Hence Batcher's algorithm can be used to sort any sequence of elements by successively merging larger and larger bitonic sequences. Given $n = 2^k$ unsorted elements, a network with $k(k+1)/2$ levels suffices. Each level contains $n/2 = 2^{k-1}$ comparators. Hence the total number of comparators is $2^{k-2}k(k+1)$. The parallel execution of each level requires constant time. Note that $k(k+1)/2 = \log n(\log n + 1)/2$. Hence the algorithm has complexity $\Theta(\log^2 n)$. ∎

Figure 4-4 is an example of bitonic merge sorting a set of eight elements. The boxes marked with a plus represent comparators that put the smaller value above the larger value. The boxes marked with a minus are comparators that put the larger value above the smaller value.

Two additional figures may help you gain a better intuition for how a series

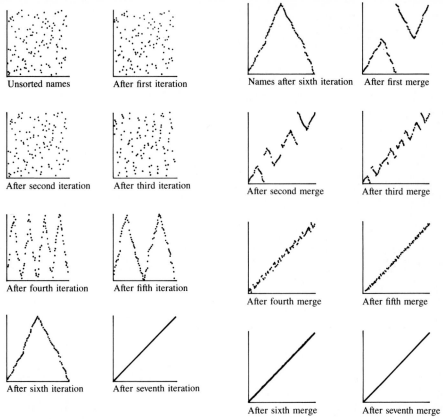

Figure 4-5 Iterations of bitonic merge-sort. The list has 128 elements; hence the sort requires log 128 = 7 iterations. Iteration i has i compare-exchange steps, for $1 \leq i \leq 7$.

Figure 4-6 Phases of the last iteration of bitonic mergesort. The list has 128 elements: hence the last iteration requires log 128 = 7 compare-exchange steps.

of bitonic merges sorts a list. Look at Figure 4-5. Each graph represents a list at some stage of the sort. The x axis represents the position of an element in the list. The y axis represents the value of an element. Hence unsorted elements form a "cloud"—see the first graph. The sorted elements form a diagonal line—see the final graph. The intermediate graphs show the form of the list after each of log n iterations. In this case $n = 128$, and log $n = 7$. Figure 4-6 illustrates what happens on the seventh, and final, iteration. A bitonic merge transforms a bitonic sequence into a set of two bitonic sequences with the property that every element in the first sequence is smaller than any element in the second sequence. Thus, once the entire set of n elements has been transformed into a single bitonic sequence, log n bitonic merges of shorter and shorter bitonic sequences are enough to complete the sort.

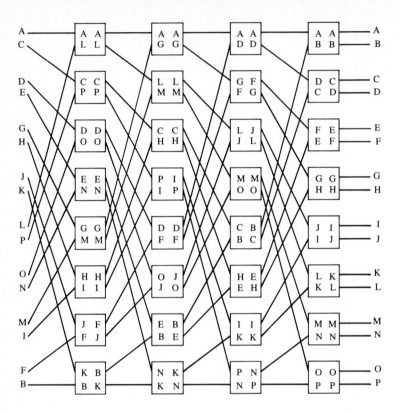

Figure 4-7 Sorting a bitonic merge sequence of length 16 by using Stone's perfect shuffle.

Theorem 4-2 (See Stone [1971].) A list of $n = 2^k$ unsorted elements can be sorted in time $\Theta(\log^2 n)$ with a network of $2^{k-1}[k(k-1)+1]$ comparators using the perfect shuffle interconnection scheme exclusively.

Stone realized that Batcher's bitonic sorter always compares elements whose indices differ by exactly 1 bit in their binary representations. Recall that the perfect shuffle routes the element at position i to the position found by cyclically rotating the binary representation of i one bit to the left. Hence two indices whose binary representations differ by exactly 1 bit can be routed to the same comparator by performing a suitable number of shuffles. Figure 4-7 shows how bitonic merge can be implemented by using the perfect shuffle interconnection scheme exclusively. Contrast this figure with Figure 4-3, in which the connections between comparators vary from stage to stage. An entire sort can be accomplished with the perfect shuffle interconnection. A sort of eight elements appears in Figure 4-8. Both algorithms require k bitonic merges to sort 2^k elements, but while the ith merge of Batcher's algorithm requires i steps, for a total of $k(k+1)/2$ steps, the 2d through kth iterations of Stone's algorithm require

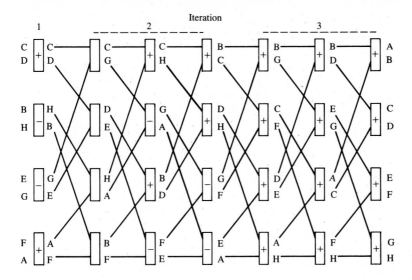

Figure 4-8 Bitonic mergesort of an unsorted list of eight elements, by using Stone's perfect shuffle interconnection.

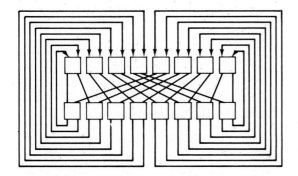

Figure 4-9 Sorting machine based upon perfect shuffle connection. (Sedgewick [1983].)

k steps, for a total of $k(k-1)+1$ steps. For a sort of eight elements there is one extra step in iteration 2, corresponding to the vertical tier of blank boxes in Figure 4-8. The blank boxes do not perform a compare-exchange operation; they output the values in the same order as they were input. These boxes are used when a number of shuffles are required before the elements to be compared are routed into the same comparator. Note that since the connections between the comparators are the same from step to step, only a single tier of comparators needs to be built (Figure 4-9).

Bitonic merge seems unsuitable for implementation in VLSI, because of the large number of path crossings. However, the efficiency of the bitonic merge has made it a popular basis for algorithms on processor array models.

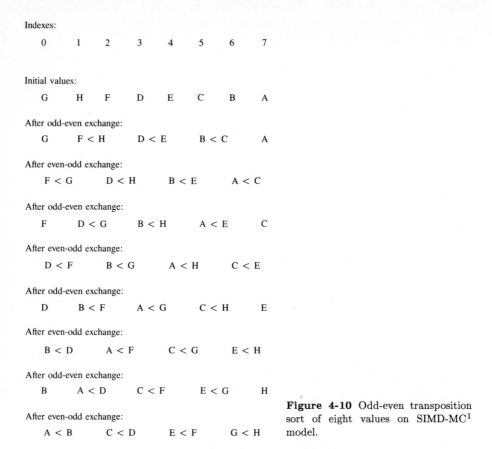

Figure 4-10 Odd-even transposition sort of eight values on SIMD-MC[1] model.

4-2 SORTING ON PROCESSOR ARRAYS

Sorting on the SIMD-MC[1] Model

The odd-even transposition sort is designed for the SIMD-MC[1] model, in which the processing elements are organized into a one-dimensional array. Assume that $A = (a_0, a_1, \ldots, a_{n-1})$ is the set of n elements to be sorted. Arrays $B = (b_0, b_1, \ldots, b_{n-1})$ and $T = (t_0, t_1, \ldots, t_{n-1})$ contain temporary values. Assume that n is even and that for all i, $0 \leq i \leq n - 1$, processor P_i contains array elements a_i, b_i, and t_i.

The algorithm requires $n/2$ iterations. Each iteration has two phases. In the first phase, called *odd-even exchange*, the value of a_i in every odd-numbered processor i (except processor $n - 1$) is compared with the value of a_{i+1} stored in even-numbered processor $i + 1$. The values are exchanged, if necessary, so that the lower-numbered processor contains the smaller value. In the second phase, called *even-odd exchange*, the value of a_i in every even-numbered processor i is compared with the value of a_{i+1} in processor $i + 1$. As in the first phase, the values are exchanged, if necessary, so that the lower-numbered processor contains the smaller value. After $n/2$ iterations the values are sorted. An example of this algorithm is given in Figure 4-10.

Theorem 4-3 (See Habermann [1972].) The complexity of sorting n elements on an SIMD-MC^1 model with n processors using odd-even transposition sort is $\Theta(n)$.

Summary of Proof. The correctness proof is based upon the fact that after i iterations of the outer **for** loop, no element can be farther than $n - 2i$ positions away from its final, sorted position. Hence $n/2$ iterations are sufficient to sort the elements, and the time complexity of the parallel algorithm is $\Theta(n)$, given n processing elements. ∎

The parallel algorithm follows.

```
ODD-EVEN TRANSPOSITION SORT (SIMD-MC¹):
begin
    for i ← 1 to n/2 do
        for all Pⱼ, 0 ≤ j ≤ n − 1 do
            if j < n − 1 and odd(j) then
                                            {Odd-even exchange}
                tⱼ ⇐ aⱼ₊₁                   {Get value from adjacent processor}
                bⱼ ← max(aⱼ,tⱼ)
                aⱼ ← min(aⱼ,tⱼ)             {Keep smaller value}
            endif
            if even(j) then
                if j > 0 then
                    aⱼ ⇐ bⱼ₋₁               {Retrieve larger value}
                endif
                                            {Even-odd exchange}
                tⱼ ⇐ aⱼ₊₁                   {Get value from adjacent processor}
                bⱼ ← max(aⱼ,tⱼ)
                aⱼ ← min(aⱼ,tⱼ)             {Keep smaller value}
            endif
            if odd(j) then
                aⱼ ⇐ bⱼ₋₁                   {Retrieve larger value}
            endif
        endfor
    endfor
end
```

Sorting on the SIMD-MC^2 Model

Theorem 4-4 Assume that $n \times n$ elements are to be sorted on the SIMD-MC^2 model. Also assume that before and after the sort the elements are distributed evenly, one element per processing element. Finally, assume that simul-

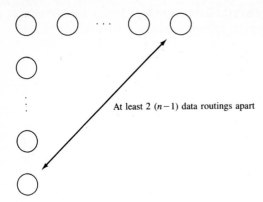

At least 2 $(n-1)$ data routings apart

Figure 4-11 The distance between names stored in opposite corners of the mesh puts a lower bound on the complexity of sorting on the SIMD-MC^2 model.

taneous data routings must always be in the same direction (east, west, north, or south). Then a lower bound on any sorting algorithm is $\Omega(n)$.

Proof. A sort may have to swap the elements originally stored in processing elements at diagonally opposite corners of the mesh. Hence a lower bound on the number of data routings needed in the worst case is $4(n-1)$ (Figure 4-11). Thus, any algorithm to sort n^2 elements on the SIMD-MC^2 model has time complexity $\Omega(n)$. Note that this implies a lower bound of $\Omega(\sqrt{n})$ to sort n elements. ∎

Theorem 4-5 (See Thompson and Kung [1977].) Given assumptions identical to those of Theorem 4-4, an algorithm exists to sort $N = n^2 = 2^k$ elements on the SIMD-MC^2 model in time $\Theta(\sqrt{N}) = \Theta(n)$.

Summary of Algorithm. The algorithm is an adaptation of Batcher's bitonic merge to the mesh. Given $N = 2^k$ elements, bitonic merge sort consists of k iterations, where each iteration i has i compare-exchange steps. Each compare-exchange requires two data routings: the first routing brings together the elements to be compared, and the second routing redistributes them. Figure 4-12 is an illustration of a network based on bitonic merge that sorts 16 elements [Knuth 1973]. Each row represents the position of an element. Arrows represent compare-exchanges. To perform a compare-exchange, the element at the position marked by the tail of the arrow is routed to the position marked by the arrowhead. After the two elements are compared, the smaller is routed back to the tail position.

Note that elements in positions whose representations differ in their least significant bit are compared every iteration, while elements in positions whose representations differ in their most significant bit are compared only on the last iteration. An efficient implementation of bitonic merge on the SIMD-MC^2 must have the property that if bit i is less significant than bit j, then a compare-exchange on bit i cannot require more data routings than a compare-exchange

Stage 1 Stage 2   Stage 3   Stage 4

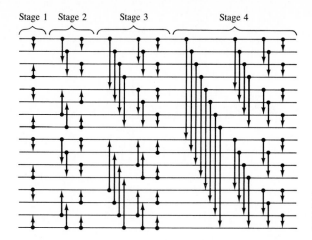

Figure 4-12 A sorting network based on bitonic merge. (Knuth [1973].)

on bit j. One way to satisfy this condition is to use a "shuffled row-major" addressing scheme, illustrated in Figure 4-13a for a 4×4 mesh. The advantage of this scheme is that "shuffling" operations occur on square subsections of the mesh, reducing the number of routing operations.

The direction of routing necessary at each compare-exchange step depends upon the index of the particular processor. Figure 4-14 illustrates the bitonic merge sort of 16 elements on a 4×4 mesh. In general, to sort $N = n^2 = 2^k$ elements by using this algorithm requires $\log N$ phases. The total number of routing steps performed is $\sum_{i=1}^{\log N} \sum_{j=1}^{i} 2^{\lfloor (j-1)/2 \rfloor}$, which is $\Theta(\sqrt{N})$. The total number of comparison steps is $\sum_{i=1}^{\log N} i$, which is $\Theta(\log^2 N)$. Thus the worst-case time complexity of bitonic merge on the SIMD-MC^2 model is $\Theta(\sqrt{N})$, making it an optimal algorithm for this model. ■

The function mapping list elements into the mesh-connected network is called the **index function**. The bitonic merge algorithm just presented sorts the list based on the shuffled-row order. What if the sorted list must be arranged in the mesh in row order (Figure 4-13b) or snakelike order (Figure 4-13c)? The following theorem shows that if each processing element has enough memory, then the sorted list can be quickly rearranged to the desired order.

Theorem 4-6 (See Thompson and Kung [1977].) If $N = n \times n$ elements have already been sorted with respect to some index function, and if each processing element can store n elements, then the N elements can be sorted with respect to any other index function in $\Theta(\sqrt{N})$ time.

Random Access Read/Write. Random access read and random access write are two frequently used data movement operations on mesh-connected models. Each operation involves two sets of processing elements: the source processors

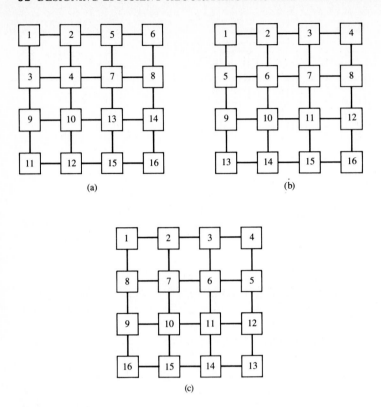

Figure 4-13 Three index functions mapping list elements into a two-dimensional mesh. (a) Shuffled row-major order. (b) Row-major order. (c) Snakelike row-major order.

and the destination processors. Source processors send a record containing a key and its associated data. Destination processors receive a record. A processing element may be both a source and a destination.

In a random access read operation, each destination processing element specifies the key of the record it wishes to receive, or else it specifies a null key, meaning it does not wish to receive a record. Several destination processing elements can specify the same key. In a random access write operation, each source processing element sends a record to a specified destination processing element, or else it specifies a null address, meaning it does not wish to send a record. Several source processing elements may specify the same address, in which case the record actually received by the destination processing element is determined by some constant-time commutative, associate, binary operation. For example, the record kept could be the one with the minimum key.

Theorem 4-7 (See Nassimi and Sahni [1981].) Random access read and random access write can be performed in $\Theta(\sqrt{n})$ time on an SIMD-MC^2 model with $\sqrt{n} \times \sqrt{n}$ processing elements.

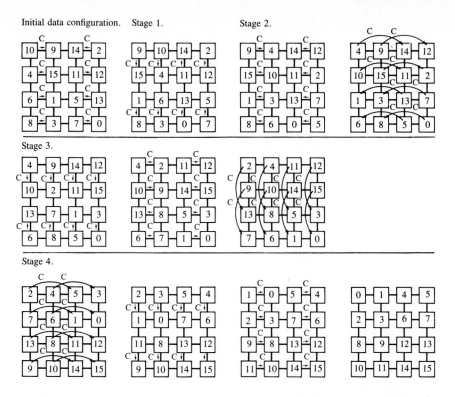

Figure 4-14 Sorting values into shuffled row-major order on the SIMD-MC^2 model. (Thompson and Kung [1977]. Copyright © 1986 Association for Computing Machinery. Reprinted by permission.)

The random access read and random access write algorithms of Nassimi and Sahni begin by sorting the records by their destinations. As we have seen, this step can be completed in $\Theta(\sqrt{n})$ time [Thompson and Kung 1977].

Sorting on the SIMD-CC Model

Bitonic merge always compares elements whose indices differ in exactly 1 bit. Since processors in the SIMD-CC model are connected if their indices differ in exactly 1 bit, it is easy to implement bitonic merge on this model. Processors replace comparators; instead of routing pairs of elements to comparators, processors route data to adjacent processors, where the elements are compared. Assume n elements are to be sorted, where $n = 2^m$ for some positive integer m. The parallel algorithm follows.

```
BITONIC MERGE SORT (SIMD-CC):
begin
    for i ← 0 to m − 1 do
        for j ← i downto 0 do
            d ← 2^j
```

```
        for all P_k where 0 ≤ k ≤ 2^m - 1 do
            if k mod 2d < d then
                t_k ⇐ a_{k+d}                    {Get value from adjacent processor}
                if k mod 2^{i+2} < 2^{i+1} then
                    b_k ← max (t_k, a_k)         {Sort low to high}
                    a_k ← min (t_k, a_k)
                else
                    b_k ← min (t_k, a_k)         {Sort high to low}
                    a_k ← max (t_k, a_k)
                endif
            endif
            if k mod 2d ≥ d then
                a_k ⇐ b_{k-d}                    {Get sorted value back}
            endif
        endfor
    endfor
  endfor
end
```

Clearly the complexity of the parallel **for all** statement is $\Theta(1)$. Hence the complexity of this parallel algorithm is $\Theta(m^2) = \Theta(\log^2 n)$, given $n = 2^m$ processors.

Sorting on the SIMD-PS Model

Not surprisingly, implementing Stone's variant of Batcher's bitonic merge algorithm on the SIMD-PS model is relatively straightforward. Instead of using comparators, elements are shuffled to a pair of processors connected with an "exchange" link. (Recall that the SIMD-PS model has "exchange" connections as well as "shuffle" connections. The exchange connections link processors whose bit representations differ in the least significant bit.) The only tricky part of the algorithm is determining whether a particular pair of elements being compared is to be sorted low to high or high to low. Stone's algorithm, which appears below, uses a mask vector M to indicate the kind of sort to be done by a particular processing element. A value of 0 corresponds to a plus comparator; a value of 1 corresponds to a minus comparator.

```
BITONIC MERGE SORT (SIMD-PS):
begin
    {Compute initial value of the mask M}
    for all P_i where 0 ≤ i ≤ n - 1 do
        r_i ← i modulo 2
        m_i ← r_i
    endfor
    for i ← 1 to m do
        for all P_i where 0 ≤ i ≤ n - 1 do
            m_i ← m_i ⊕ r_i    {Exclusive OR}
            shuffle (m_i)
        endfor
    endfor
```

```
COMPARE-EXCHANGE (A, M)
for i ← 1 to m − 1 do
    for all P_i where 0 ≤ i ≤ n − 1 do
        shuffle (r_i)
        m_i ← m_i ⊕ r_i   {Exclusive OR}
        for j ← 1 to m − 1 − i do
            shuffle (a_i)
            shuffle (m_i)
        endfor
    endfor
    for j ← m − i to m do
        for all P_i where 0 ≤ i ≤ n − 1 do
            shuffle (a_i)
            shuffle (m_i)
        endfor
        COMPARE-EXCHANGE (A, M)
    endfor
endfor
end
```

```
COMPARE-EXCHANGE (A, M):
begin
    for all P_i where 0 ≤ i ≤ n − 1 do
        if even(i) then
            t_i ⇐ a_{i+1}
            if m_i = 0 then        {Sort low to high}
                b_i ← max(a_i, t_i)
                a_i ← min(a_i, t_i)
            else                   {Sort high to low}
                b_i ← min(a_i, t_i)
                a_i ← max(a_i, t_i)
            endif
        endif
        if odd(i) then
            a_i ⇐ b_{i−1}
        endif
    endfor
end
```

This algorithm requires $m(m+1)/2$ compare-exchange steps, $m(m-1)$ shuffle steps of the vector A, and $2m-1$ shuffles of the vectors M and R. Since $m = \log n$, the time complexity of this algorithm is $\Theta(\log^2 n)$ with n processors.

4-3 A MULTIPROCESSOR IMPLEMENTATION OF QUICKSORT

Quicksort is a commonly used sorting algorithm on serial computers. Its popularity is due to its asymptotically optimal average-case behavior: Given a random permutation of n elements, quicksort can be expected to sort the elements in about $2n \log n$ comparisons [Baase 1978].

Quicksort is a recursive algorithm that repeatedly divides an unsorted sublist into two smaller sublists and a supposed median value. One of the smaller sublists contains values less than the median value; the other sublist contains values greater than the median. The median value, located between the two smaller sublists, is in its correctly sorted position, since all the values to the left are smaller and all the values to the right are larger. Given an initially unsorted list, then, the quicksort algorithm chooses one element as the supposed median (e.g., the first element). After a single partitioning step, the list is divided into two sublists, and the algorithm recursively partitions each of the sublists. Quicksort is an example of an algorithm using the divide-and-conquer approach: Once a list has been partitioned, the two unsorted sublists form independent problems that can be solved simultaneously.

Consider the following parallel quicksort algorithm. A number of identical processes, one per processor, execute the parallel algorithm. The elements to be sorted are stored in an array in global memory. A stack in global memory stores the indices of subarrays that are still unsorted. When a process is without work, it attempts to pop the indices for an unsorted subarray off the global stack. If it is successful, the processor partitions the subarray, based on a supposed median element, into two smaller arrays, containing elements less than or greater than the supposed median value. After the partitioning step, which is identical to the partitioning step performed by the serial quicksort algorithm, the process pushes the indices for one of the subarrays onto the global stack of unsorted subarrays and repeats the partitioning process on the other subarray.

What speedup can be expected from this parallel quicksort algorithm? Note that it takes $k-1$ comparisons to partition a subarray containing k elements. The expected speedup is computed by assuming that one comparison takes 1 unit of time and finding the ratio of the expected number of comparisons performed by the sequential algorithm to the expected time required by the parallel algorithm. To simplify the analysis, assume that $n = 2^k - 1$ and $p = 2^m$, where $m < k$. Also assume that the supposed median is always the true median, so that each partitioning step always divides an unsorted subarray into two subarrays of equal size.

With these assumptions the number of comparisons made by the sequential algorithm can be determined by solving the following recurrence relation:

$$T(n) = \begin{cases} n - 1 + 2T\dfrac{n-1}{2} & \text{for } n = 7, 15, 31, \dots \\ 2 & \text{for } n = 3 \end{cases}$$

The solution to this recurrence relation is

$$T(n) = (n+1)\log(n+1) - 2n$$

The parallel algorithm has two phases. In the first phase there are more processes than arrays to be sorted. For example, to begin there is only a single unsorted array. All but one of the processes must wait while a single process partitions the array. This iteration, then, requires $n - 1$ time units to perform $n - 1$ comparisons. If we assume $p \geq 2$, two processes can partition the two resulting subarrays in $(n-1)/2-1$ time units, performing $n-3$ comparisons. Similarly, if $p \geq 4$, the third iteration requires time at least $[(n-1)/2-1]/2-1$ to perform $n-7$ comparisons. For the first $\log p$ iterations, then, there are at least as many processes as partitions, and the time required by this phase of the parallel quicksort algorithm is

$$T_1(n) = 2(n+1)\left(1 - \frac{1}{p}\right) - \log p$$

The number of comparisons performed is

$$C_1(n) = (n+1)\log p - 2(p-1)$$

In the second phase of the parallel algorithm there are more subarrays to be sorted than processes. All the processes are active. If we assume that every process performs an equal share of the comparisons, then the time required is simply the number of comparisons performed divided by p. Hence

$$C_2(n) = T(n) - C_1(n)$$

$$T_2(n) = \frac{C_2(n)}{p}$$

The estimated speedup achievable by the parallel quicksort algorithm is the sequential time divided by the parallel time:

$$\text{Speedup} = \frac{T(n)}{T_1(n) + T_2(n)}$$

For example, the best speedup we could expect with $n = 1023$ and $p = 8$ is approximately 3.375. Why is this limit so low? The problem with quicksort is its divide-and-conquer nature. Until the first subarray is partitioned, there are no more partitionings to do. Even after the first partitioning step has completed, there are only two subarrays to work with. Hence many processes spend a lot of time at the beginning of the parallel algorithm's execution, waiting for work.

In 1978 Raskin implemented parallel quicksort on a 10-processor Cm*; Deminet adapted it to the 50-processor Cm* in 1982, "modifying the algorithm to cut down on references to the shared stack" [Gehringer, Jones, and Segall 1982]. Figure 4-15 illustrates the speedup of the parallel quicksort algorithm on Cm* for various numbers of processes (processors), given an array of 20,480 elements.

The speedup of parallel quicksort on a single-PEM Denelcor HEP running under the HEP/UPX operating system is shown in Figure 4-16.

4-4 FAST FOURIER TRANSFORM

The Fourier transform has many uses in science and engineering. Applications include speech transmission, coding theory, and image processing. Hence the discovery of a fast algorithm to perform the discrete Fourier transform was of

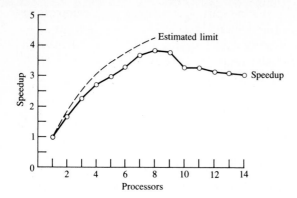

Figure 4-15 Speedup of parallel quicksort algorithm on Cm*.

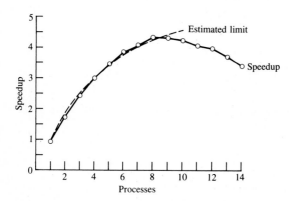

Figure 4-16 Speedup of parallel quicksort algorithm on single-PEM Denelcor HEP.

particular significance. This section presents the sequential fast Fourier transform algorithm and its straightforward parallelization on the SIMD-CC model.

To benefit the reader unfamiliar with the fast Fourier transform, we begin with a motivating example. The product of the two $(n-1)$-degree polynomials

$$p(x) = \sum_{i=0}^{n-1} a_i x^i \quad \text{and} \quad q(x) = \sum_{i=0}^{n-1} b_i x^i$$

is the $(2n-2)$-degree polynomial

$$p(x)q(x) = \sum_{i=0}^{2n-2} \sum_{j=0}^{i} a_j b_{i-j} x^i$$

The coefficients of the resulting polynomial $p(x)q(x)$ can be found by performing the above summation, convoluting the coefficient vectors of the original polynomials. By using this algorithm, polynomial multiplication can be performed in $O(n^2)$ time.

The observation that a polynomial $p(x)$ of degree $n-1$ is uniquely determined by its values at the nth roots of unity leads to another algorithm. Evaluating

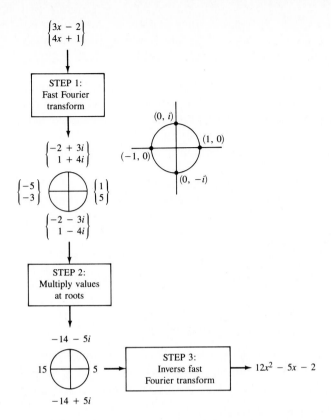

Figure 4-17 Multiplication of two polynomials by using the fast Fourier transform algorithm is a three-step process.

the Fourier transform of the coefficients of a polynomial of degree $n-1$ generates the value of the polynomial at the n roots of unity. The value of the product at any point is simply the product of the values of the two polynomials at that point. Conversely, performing the inverse Fourier transform on the values of a polynomial of degree $n-1$ at the n roots of unity generates the coefficients of that polynomial.

The following polynomial multiplication algorithm results. Consider two polynomials of degree $n-1$. The product polynomial will have degree $2n-2$. Let N be the smallest power of 2 such that $N \geq 2n-1$, the number of coefficients in the product. First, the Fourier transform is used to evaluate the two polynomials at the N roots of unity. Second, the two values at each root are multiplied. Third, the inverse Fourier transform takes the resultant values and generates the coefficients of the product polynomial.

Figure 4-17 illustrates this algorithm. The two polynomials to be multiplied are $3x-2$ and $4x+1$. The Fourier transform is applied to these polynomials to determine their values at the fourth roots of unity. The fourth roots of unity are

$\{1, i, -1, \text{and} -i\}$, where $i = \sqrt{-1}$. The pair of values at each root are multiplied. For example, $(-2 + 3i) \times (1 + 4i) = -14 - 5i$. The four resulting values uniquely determine the product polynomial, whose coefficients can be found by applying the inverse Fourier transform.

Why would we want to use such an involved algorithm to multiply polynomials? The reason is that a "fast" algorithm to perform the discrete Fourier transform exists. The fast Fourier transform (FFT) uses a divide-and-conquer technique to perform the discrete Fourier transformation in time $O(n \log n)$. Hence two polynomials of degree $n - 1$ can be multiplied in time $O(n \log n)$ by using the FFT.

Before we present the FFT algorithm, four functions must be defined. Function $\mathsf{BIT\_0}(i, j)$ returns the integer found by setting the bit corresponding to 2^j in i to 0. For example,

$$\mathsf{BIT\_0}(6, 1) = \mathsf{BIT\_0}((110)_2, 1) = (100)_2 = 4$$
$$\mathsf{BIT\_0}(11, 2) = \mathsf{BIT\_0}((1011)_2, 2) = (1011)_2 = 11$$

$\mathsf{BIT\_1}(i, j)$ returns the integer found by setting the bit corresponding to 2^j in i to 1. For example,

$$\mathsf{BIT\_1}(3, 2) = \mathsf{BIT\_1}((011)_2, 2) = (111)_2 = 7$$
$$\mathsf{BIT\_1}(9, 0) = \mathsf{BIT\_1}((1001)_2, 0) = (1001)_2 = 9$$

The function $\mathsf{OMEGA}(i, j, k)$ returns the lth primitive root of unity, where l is the value found by reversing the $i + 1$ most significant bits of the k-bit integer j and then padding the result with $k - i - 1$ zeros. For example,

$$\mathsf{OMEGA}(1, 5, 4) = \mathsf{OMEGA}(1, (0101)_2, 4) = \omega^{(1000)_2} = \omega^{16}$$

$$\mathsf{OMEGA}(2, 5, 4) = \mathsf{OMEGA}(2, (0101)_2, 4) = \omega^{(0100)_2} = \omega^4$$

Function $\mathsf{REVERSE}(i, j)$ returns a value between 0 and $2^j - 1$, found by reversing the bit representation of the j-bit integer i. For example,

$$\mathsf{REVERSE}(4, 3) = \mathsf{REVERSE}((100)_2, 3) = (001)_2 = 1$$
$$\mathsf{REVERSE}(7, 5) = \mathsf{REVERSE}((00111)_2, 5) = (11100)_2 = 28$$

The fast Fourier transform algorithm, presented below, transforms coefficient vector a into value vector b. To make the algorithm easier to read, parentheses have been used instead of subscripts.

```
FAST FOURIER TRANSFORM (SISD):
begin
    for i ← 0 to 2^k − 1 do
        r(i) ← a(i)
    endfor
    for i ← 0 to k − 1 do
        for j ← 0 to 2^k − 1 do
            s(i) ← r(i)
        endfor
```

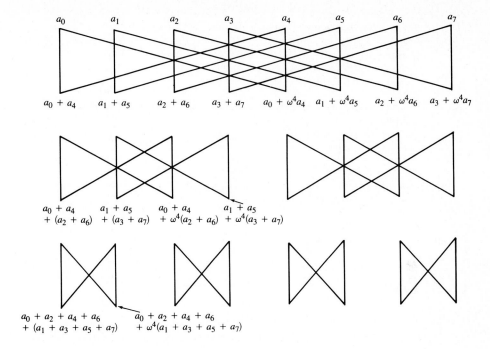

a_0 a_1 a_2 a_3 a_4 a_5 a_6 a_7

$a_0 + a_4$ $a_1 + a_5$ $a_2 + a_6$ $a_3 + a_7$ $a_0 + \omega^4 a_4$ $a_1 + \omega^4 a_5$ $a_2 + \omega^4 a_6$ $a_3 + \omega^4 a_7$

$\begin{aligned} &a_0 + a_4 \\ &+ (a_2 + a_6) \end{aligned}$ $\begin{aligned} &a_1 + a_5 \\ &+ (a_3 + a_7) \end{aligned}$ $\begin{aligned} &a_0 + a_4 \\ &+ \omega^4 (a_2 + a_6) \end{aligned}$ $\begin{aligned} &a_1 + a_5 \\ &+ \omega^4 (a_3 + a_7) \end{aligned}$

$\begin{aligned} &a_0 + a_2 + a_4 + a_6 \\ &+ (a_1 + a_3 + a_5 + a_7) \end{aligned}$ $\begin{aligned} &a_0 + a_2 + a_4 + a_6 \\ &+ \omega^4 (a_1 + a_3 + a_5 + a_7) \end{aligned}$

Figure 4-18 Computation of the fast Fourier transform. (Aho, Hopcroft, and Ullman [1974].)

```
    for j ← 0 to 2^k − 1 do
        r(i) ← s(BIT_0(j, k − 1 − i)) + OMEGA(i, j, k) × s(BIT_1(j, k − 1 − i))
    endfor
endfor
for i ← 0 to 2^k − 1 do
    b_i ← r(REVERSE(i, k))
endfor
end
```

Figure 4-18 illustrates how the coefficients of the original polynomial are combined in the FFT algorithm. You should recognize the pattern immediately: it is identical to a butterfly network. Hence this algorithm is easily parallelized on a cube-connected processor array network. A parallel FFT algorithm for the SIMD-CC model follows.

```
FAST FOURIER TRANSFORM (SIMD-CC with p = 2^k)
begin
    for all P(i), where 0 ≤ i ≤ 2^k − 1 do
        r(i) ← a(i)
    endfor
```

```
for i ← 0 to k − 1 do
    for all P(j), where 0 ≤ j ≤ 2^k − 1 do
        s(j) ← r(j)
        adjacent(j) ← BIT_COMPLEMENT(j, k − 1 − i)
        if adjacent(j) = BIT_0(j, k − 1 − i) then
            s(j) ⇐ adjacent(j)
        else
            s(j) ← r(j)
        endif
        if adjacent(j) = BIT_1(j, k − 1 − i) then
            t(j) ⇐ r(adjacent(j))
        else
            t(j) ← r(j)
        endif
        r(j) ← s(j) + OMEGA(i, j, k) × t(j)
    endfor
endfor
end
```

4-5 SUMMARY

Sorting is an important utility on both serial and parallel computers. It is the foundation of many important parallel algorithms. On mesh-connected and pyramid models, for example, random access read and write are expressed in terms of sorting.

A wide variety of parallel sorting algorithms have been proposed. This chapter has described only a few. Batcher's bitonic merge algorithm, although not directly implementable in VLSI, has been the basis of sorting algorithms for the SIMD-PS, SIMD-CC, and even multicomputer models. Sorting can be performed in $\Theta(\log^2 n)$ time on the SIMD-PS and SIMD-CC models, given n processing elements. The SIMD-MC model with $q = 2$ cannot sort n elements in $O(\log^k n)$ time; data routing puts a lower bound of $\Omega(\sqrt{n})$ for sorting on this model.

The development of a parallel quicksort algorithm for MIMD computers shows that algorithms relying on a divide-and-conquer approach may yield unsatisfactory speedups on multiprocessors, if they do not provide enough work for the processors during all phases of the program's execution.

Sorting and the fast Fourier transform share the characteristic that no result can be computed without examining every input. Hence every processor needs to inspect all the data, either directly or indirectly. For this reason these algorithms are most efficiently implemented on SIMD architectures that allow data to be transferred efficiently across the network.

Note that hypercubic multicomputers share the same processor organization as cube-connected processor arrays. Thus the SIMD-CC algorithms presented in this chapter would map nicely onto hypercubes. The efficiency of these algorithms would be heavily dependent upon the communication delay of the particular hypercube being used.

BIBLIOGRAPHIC NOTES

Parallel sorting algorithms have been the object of much study. In fact, an entire book has been devoted to the topic: *Parallel Sorting Algorithms*, by S. G. Akl [Akl 1985]. Two survey articles on parallel sorting are by Lakshmivarahan, Dhall, and Miller [1984]; and Bitton et al. [1984]. Knuth [1973] also discusses parallel sorting algorithms, including odd-even transposition sort.

Atjai, Komlós, and Szemerédi [1983] describe an algorithm that uses $O(n \log n)$ processors to sort n elements in time $O(\log n)$. Leighton [1984] shows how this algorithm can be used to sort n elements in time $O(\log n)$ by using n processors, which is asymptotically optimal. However, Leighton points out that the constant of proportionality of this algorithm is immense and that unless $n > 10^{100}$, these algorithms would be slower in practice than other parallel sorting algorithms.

Other references to parallel sorting algorithms based on bitonic merge include Baudet and Stevenson [1978]; Brock, Brooks, and Sullivan [1981]; Flanders [1982]; Kumar and Hirschberg [1983]; Lorin [1975]; Meertens [1979]; Nassimi and Sahni [1979, 1982]; Perl [1983]; Preparata [1978]; Preparata and Vuillemin [1981]; Rudolph [1984]; and Schwartz [1980]. In addition to the bitonic merge sort discussed in this chapter, Thompson and Kung [1977] present another sorting algorithm for the SIMD-MC^2 model, called the s^2-way odd-even merge, which also can sort $N = n^2$ elements in time $\Theta(\sqrt{N})$.

Extensions to Stone's perfect shuffle implementation of bitonic merge include Brock, Brooks, and Sullivan [1981]; Hoey and Leiserson [1980]; Kleitman et al. [1981]; Knuth [1973]; Leighton [1983]; Meertens [1979]; Schwartz [1980]; and Stone [1978].

Early references to odd-even transposition sort include Demuth [1956], Knuth [1973], and Kung [1980]. Besides the references cited earlier in the chapter, other implementations of the odd-even transposition sort include Chen et al. [1978a, 1978b]; Kramer and van Leeuwen [1982]; Kumar and Hirschberg [1983]; Lee, Chang, and Wong [1981]; and Miranker, Tang, and Wong [1983]. Baudet and Stevenson [1978] generalized the algorithm to the case where each processor sorts a subsequence, rather than a single value.

Thompson and Kung [1977] discuss the row-major, snakelike row-major, and shuffled row-major processor numbering schemes on the SIMD-MC^2 model. Nassimi and Sahni [1979] discuss the lower bound for sorting on the SIMD-MC^2 model. They also describe algorithms to sort elements into row-major order and snakelike row-major order.

Muller and Preparata [1975] first proposed the idea of sorting by using a tree of processors to augment a mesh-connected processor array. Leighton [1981] derives lower bounds for several computations performed on this model. Other references to sorting algorithms based on treelike networks of processors include Aggarwal [1984], Bentley and Kung [1979], Horowitz and Zorat [1983], Mead and Conway [1980], Stout [1983a], and Tanimoto [1982a, 1982b]. Implementing tree machines in VLSI is the subject of papers by Bhatt and Leiserson [1982],

Leiserson [1980], Mead and Rem [1979], Ruzzo and Snyder [1981], and Valiant [1981].

Todd [1978] describes how to perform mergesort on a pipeline of processors.

An **enumeration sort** finds the sorted position of each elements by determining the number of elements smaller than it. Muller and Preparata [1975] proposed a parallel enumeration sort. Other implementations of enumeration sorts appear in Hsiao and Snyder [1983]; Leighton [1981]; Nath, Maheshwari, and Bhatt [1983]; and Yasuura, Tagaki, and Yajima [1982].

A randomized parallel sorting algorithm has been designed by Reif and Valiant [1983].

Valiant [1975], Hirschberg [1978], Horowitz and Zorat [1983], Kruskal [1983], Preparata [1978], Reischuk [1981], and Shiloach and Vishkin [1981] have proposed parallel sorting algorithms for shared-memory SIMD models.

Chabbar [1980] presents a parallel enumeration sort for the MIMD model. Besides the earlier citations, other references for MIMD quicksort algorithms include Lorin [1975], Quinn [1987], and Robinson [1977]. Quinn [1987], Robinson [1977], and Tolub and Wallach [1978] have described parallel sorting algorithms for MIMD computers based on merging. Tolub and Wallach [1978] also discuss a parallel bucket sort. Quinn [1987] discusses parallel shellsort.

Loui [1984] derives upper and lower bounds for sorting on distributed computers. References to distributed sorting algorithms include Rotem, Santoro, and Sidney [1983] and Wegner [1982].

Other sorting networks are discussed in Armstrong and Rem [1982]; Atjai, Komlós, and Szemerédi [1983]; Carey, Hansen, and Thompson [1982]; Chen et al. [1978a, 1978b]; Chin and Fok [1980]; Chung, Luccio, and Wong [1980a, 1980b]; De Bruijn [1984]; Dowd et al. [1983]; Hong and Sedgewick [1982]; Lee, Chang, and Wong [1981]; Miranker, Tang, and Wong [1983]; Moravec [1979]; Mukhopadhyay [1981]; Mukhopadhyay and Ichikawa [1972]; Tseng and Lee [1984a, 1984b]; Winslow and Chow [1981, 1983]; and Wong and Ito [1984].

Parallel external sorting algorithms are discussed in Akl and Schmeck [1984]; Bonuccelli, Lodi, and Pagli [1984]; Lee, Chang, and Wong [1981]; and Yasuura, Tagaki, and Yajima [1982].

Introductions to the fast Fourier transform algorithm include Aho, Hopcroft, and Ullman [1974]; Baase [1978]; and Horowitz and Sahni [1978]. Flanders [1982] describes an implementation of the fast Fourier transform on the ICL DAP. Other articles in the literature related to parallel fast Fourier transforms include Bergland [1972]; Brigham [1973]; Chow, Vranesic, and Yen [1983]; Corinthios and Smith [1975]; Cyre and Lipovski [1972]; Dere and Sakrison [1970]; Jesshope [1980]; Kulkarni and Yen [1982]; Lambiotte and Korn [1979]; Parker [1980]; Pease [1968, 1977]; Preparata and Vuillemin [1981]; Ramamoorthy and Chang [1971]; Redinbo [1979]; Stone [1971]; Thompson [1983a, 1983b]; Wold and Despain [1984]; and Zhang and Yun [1984].

EXERCISES

4-1 Show how the following 16 values would be sorted by Batcher's bitonic merge algorithm: 7, 9, 10, 2, 3, 6, 16, 1, 14, 5, 15, 8, 4, 11, 13, 12.

4-2 In general the bitonic sort of 2^k numbers requires how many comparison steps, with each step using 2^{k-1} comparators?

4-3 Given a bitonic sequence, prove that after a single compare-exchange step the sequence has been divided into two sequences with the property that every element of the first sequence is smaller than every element of the second sequence.

*4-4 Given a bitonic sequence, prove that after a single compare-exchange step the resulting sequences are bitonic.

4-5 How many steps does Stone's bitonic sorter require for n values, where $n = 2^k$?

4-6 Use odd-even transposition sort to sort these eight values: 5, 8, 3, 2, 4, 6, 4, 1.

4-7 Sorting on the SIMD-MC^2 model has complexity $O(\sqrt{n})$. How would the complexity change if the processing elements on the left edge were connected to processing elements on the right edge and the processing elements on the top edge were connected to processing elements on the bottom edge, giving every processing element four neighbors?

4-8 Given an SIMD-MC^j model with N processors, prove that bitonic sort can be performed in time $\Theta(N^{1/j})$, using the j-way shuffled row-major index scheme.

4-9 Prove that the bound in the previous exercise is optimal.

4-10 Why does Stone's perfect shuffle algorithm have to carry around a mask vector, while the SIMD-CC algorithm does not?

4-11 A problem with the parallel quicksort algorithm presented is that too many processes wait too long before getting work to do. Find at least two ways to modify the parallel algorithm so that more processes are busy sooner.

4-12 What is the worst-case time complexity of the parallel quicksort algorithm?

4-13 Explain why it is not appropriate for the processes executing the parallel quicksort algorithm to terminate when they discover that the stack of unsorted subarrays is empty. Devise a halting condition that is both correct and efficient.

FIVE

DICTIONARY OPERATIONS

This chapter describes parallel algorithms to solve the problems of searching an ordered table for the existence of a particular key, inserting keys into an ordered table, and deleting keys. Efficient sequential algorithms have been developed to allow dictionary operations to be performed in time logarithmic in the size of the table, an enormous improvement over the linear time that would be needed if keys were kept in an unordered list. Sometimes it is important for multiple processes to perform dictionary operations concurrently. For example, it is almost inevitable that a parallel compiler would have to allow more than one process to access the symbol table simultaneously.

Search algorithms operate on elements, called **keys**, stored in a **table** of finite size. The goal is to organize the table and implement the algorithms so that inserting keys and their associated data into the table, deleting keys and data from the table, and searching for keys in the table execute as quickly as possible.

Searching is a trivial operation on SIMD computers. Searching for a particular key can be accomplished in three steps. First, the key to be searched for is broadcast to every processing element. Second, every processing element compares its keys with the desired key, setting a flag to indicate whether there is a match. Third, the existence of a *true* flag is ascertained. Hence in this chapter we concentrate on algorithms for multiprocessors.

The chapter is divided into two principal sections. Section 5-1 studies the inherent complexity of parallel search algorithms. We find that using multiple processors to perform a single operation is not particularly efficient—speedup is no more than logarithmic in the number of processors used. Section 5-2 presents two algorithms that perform **batch searching**; that is, they allow a number of searches to proceed concurrently. The first algorithm, developed by Ellis, allows concurrent search and insertion on AVL trees. The second algorithm, credited to Manber and Ladner, allows deletions as well, although the AVL property is sacrificed.

5-1 COMPLEXITY OF PARALLEL SEARCH

How quickly can the search for a single key be performed on a parallel computer? It is useful to have a bound on the number of operations required to perform a particular function on a parallel computer, because we then have a standard by which to gauge various proposed algorithms. The SIMD-SM-R model is frequently used to find this bound. Recall that the SIMD-SM-R model is unrealistic, because it assumes a conflict-free access to a common shared memory and avoids communications issues. However, since the complexity of algorithms on the SIMD-SM-R model depends solely on the number of computations, the model provides a useful lower bound on the inherent computational complexity of the algorithm.

Theorem 5-1 (See Kruskal [1982, 1983].) Given positive integers k, n, and p, where $n = (p+1)^k - 1$, searching for a key in an n-element table while using the SIMD-SM-R model requires $\leq \lceil \log(n+1)/\log(p+1) \rceil$ comparisons. This bound is tight.

Proof. We use induction on k to show that $\lceil \log(n+1)/\log(p+1) \rceil$ comparisons are sufficient.

Basis: Let $k = 1$. Then $n = (p+1)^1 - 1 = p$. Clearly one comparison step is sufficient for p processors to determine whether the key is in the table, and $\lceil \log(p+1)/\log(p+1) \rceil = 1$.

Induction: Assume true for all tables of size $(p+1)^j - 1$, where $1 \leq j < k$. To search a list of size $(p+1)^k - 1$, during the first comparison processor i, for $1 \leq i \leq p$, compares the key with the table element indexed by $i(p+1)^{k-1}$. After this step, either one of the table elements has matched the key, or else the key lies inside one of the unexamined subsections of the table. All these unexamined subsections have size $(p+1)^{k-1} - 1$. By the induction hypothesis, $k - 1$ comparison steps are sufficient to search any of these subtables (Figure 5-1).

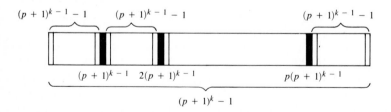

Figure 5-1 First induction step for Theorem 5-1.

The second step of the proof is to show that the bound is tight. During the first parallel comparison step, only p elements of the table are compared with the key. There must be one or more contiguous unexamined segments of the table with length at least

$$\left\lceil \frac{n-p}{p+1} \right\rceil \geq \frac{n-p}{p+1} = \frac{n+1}{p+1} - 1$$

An inductive argument shows that after k parallel comparison steps there must be one or more contiguous unexamined segments of the table with length at least

$$\frac{n+1}{(p+1)^k} - 1$$

Thus the number of steps required by any parallel algorithm in the worst case is at least the minimum k satisfying

$$n + 1/(p+1)^k - 1 \leq 0$$
$$\Rightarrow \qquad k \geq \log(n+1)/\log(p+1)$$
$$\Rightarrow \qquad k = \lceil \log(n+1)/\log(p+1) \rceil \quad \blacksquare$$

Using the results of the previous theorem, we can calculate the speedup achieved by the parallel searching algorithm. Assume that we are to search a sorted list. Taking the number of comparisons made by the sequential binary search algorithm in the worst case divided by the least number of comparisons made by the parallel algorithm in the worst case, we find that

$$S = \frac{\lceil \log(n+1) \rceil}{\lceil \log(n+1)/\log(p+1) \rceil}$$
$$\approx \log(p+1)$$

In other words, the speedup achieved through parallelization is logarithmic only in the number of processors used.

What conclusions should we draw from this theorem? Since the SIMD-SM-R model can require as many as $\lceil \log(n+1)/\log(p+1) \rceil$ comparison steps, it is safe to assume that *any* model of parallel computation can require at least $\lceil \log(n+1)/\log(p+1) \rceil$ comparison steps in the worst case. More realistic models could also have other time-consuming operations, such as data routing and process synchronization. Thus real parallel computers will experience speedup that is no more than logarithmic in the number of processors used, and it does not seem fruitful to attempt to speed up a single search. The sequential algorithm is quite fast already, having logarithmic complexity.

Our strategy for the rest of the chapter, then, is to speed up a series of searches. Search algorithms usually do not appear as ends in themselves; they are frequently called subalgorithms for larger problems. Hence it is likely that a method to perform a number of searches, insertions, and deletions in parallel would be useful.

5-2 SEARCHING ON TIGHTLY COUPLED MULTIPROCESSORS

A logical way to approach the problem is to store the search tree in a shared global memory and associate processors with individual requests. A single processor is responsible for responding to a command to insert, delete, or search.

Ellis's Algorithm

Ellis [1980b] has suggested a parallel algorithm that allows concurrent inserting and searching to take place in AVL trees. It is assumed you have a general knowledge of graph theoretic terms. To recapitulate some key definitions:

Definition 5-1 (See Aho, Hopcraft, and Ullman [1974].) The **height** of a rooted tree is the length of the longest path from the root to a leaf node. The "empty tree"—tree without even a root—has height -1.

Definition 5-2 An **AVL tree** is a binary tree having the property that for any node v in the tree, the difference in height between the left and right subtrees of node v is no more than 1.

Baer and Schwab [1977] have shown that AVL tree construction is the asymptotically optimal way of keeping binary search trees balanced when the only two operations to be performed are searching and inserting. As keys are added to the AVL tree, two types of rotations are sufficient to keep the tree balanced: single rotation and double rotation. Both rotations occur when the two subtrees of a particular node do not have the same height and the height of the taller subtree increases. These rotations, which are illustrated in Figure 5-2, require $O(\log n)$ time.

Sequential Insertion Algorithm. Each node v of an AVL tree has four fields associated with it: $key(v)$, containing a unique key; $left(v)$, a pointer to the left subtree; $right(v)$, a pointer to the right subtree; and $bal(v)$, an integer whose value is 0 if the left and right subtrees are balanced, whose value is -1 if the left tree is taller than the right subtree, and whose value is $+1$ if the right tree is taller than the left subtree.

The sequential insertion algorithm has three phases. In the first phase the tree is searched to find the appropriate place to attach the new leaf node. During the search a pointer is set to point to the last node encountered whose subtrees have different height. This node, hereafter referred to as node c, is called the **critical node**. If every node along the search path has balanced subtrees, then the root is the critical node. The first phase ends by inserting the new leaf node.

Phase 2 consists of traversing all the nodes on the path between the newly inserted node v and the critical node c. For each such node w, if $key(v) < key(w)$, then $bal(w)$ is given the value -1; otherwise, $bal(w)$ is given the value $+1$.

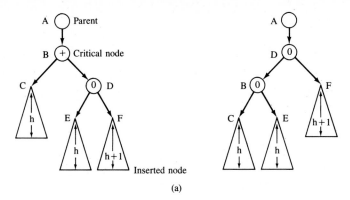

(a)

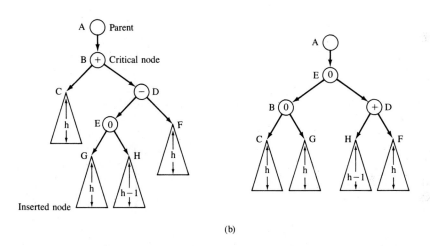

(b)

Figure 5-2 Rotations to keep AVL tree balanced. (a) Single rotation. (b) Double rotation.

Phase 3 modifies the value of $bal(c)$ and rotates the tree if necessary. If $bal(c) = +1$ and v was inserted in the left subtree, or if $bal(c) = -1$ and v was inserted in the right subtree, then $bal(c)$ is changed to 0 and no rotation is necessary. If c is the root node and $bal(c) = 0$, then $bal(c)$ is set to -1 or $+1$, depending upon whether the insertion was into the left or right subtree. Otherwise v was inserted into the subtree with greater height, and a single or double rotation must be performed.

Parallel Algorithm. The goal of the parallel algorithm is to keep as many search and insertion processes active as possible. It is not possible for each process to ignore the other processes. For example, consider the AVL tree of

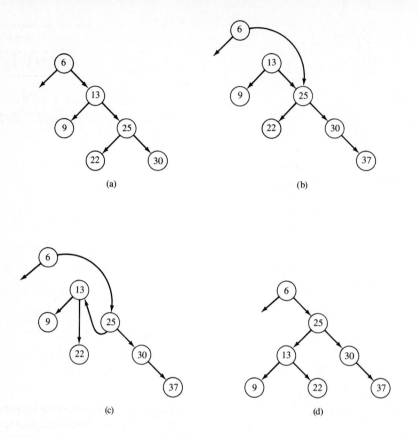

Figure 5-3 Transformations in AVL tree as value 37 is added.

Figure 5-3a. Assume that two processes are active: the first is inserting the value 37, and the second is searching for the value 13. After the value 37 is inserted into the tree, a single rotation must be performed to preserve the AVL property. This rotation requires that the values of certain *left* and *right* pointers be changed. The process of changing these pointers is illustrated in Figure 5-3b and c. (Figure 5-3d is simply a redrawing of Figure 5-3c.) Suppose that the process searching for the key 13 follows the right child of node 6 and the left child of node 25 when the tree is in the state depicted by Figure 5-3b. The value 13 would not be found, even though it is in the tree. Another contention problem could occur if two processes attempted to rotate subtrees with nodes in common.

Ellis solves contention problems by adding three lock fields to each node. These lock fields enable search processes to be locked out of a subtree being rotated. They also enable an insert process to lock subsequent insert processes out of the entire subtree rooted by the parent of the critical node.

Three lock fields are added to each node of the AVL tree: ρ lock, α lock, and ξ lock. A process performing a search must hold a node's ρ lock before examining the contents of that node. The α locks are set by an insert process to keep other insert processes off of the path from the parent of the critical node to the point of insertion. The ξ locks are used to exclude search processes from nodes involved in a rotation. More than one search process can share a single ρ lock, and multiple search processes can hold a node's ρ lock while a single insert process holds an α lock on the node. However, α locks and ξ locks may not be shared, and if an insert process holds a node's ξ lock, then no other processes can hold the α lock or the ρ lock.

In the case of a single rotation, the ξ locks of the critical node and its parent must be set. A double rotation requires that the ξ locks be set on the critical node, the parent of the critical node, and the child of the critical node lying along the insertion path. Figure 5-4 illustrates what locks are held on an AVL tree at various times during the execution of four processes performing the following insert and search operations:

Task			
Insert 51	Search for 46	Insert 25	Search for 17
Lock α 26	Lock ρ 26		
	Unlock ρ 26		
Lock α 34	Lock ρ 34		
Lock α 49	Unlock ρ 34		
	Lock ρ 49		
Figure 5-4a			
Lock α 66	Unlock ρ 49		
Insert key 51	Terminate		
Lock ξ 26			
Lock ξ 34		Waiting for α 26	
Figure 5-4b			
Rotate at 34		Lock α 26	
Release all locks		Lock α 20	
Terminate		Lock α 23	
		Insert key 25	Lock ρ 26
		Waiting for ξ 26	
Figure 5-4c			
			Unlock ρ 26
		Lock ξ 26	Lock ρ 20

During execution of Ellis's algorithm, a process performing an insertion sets α locks as it traverses the tree. The α locks from the parent of the critical node through the place of insertion remain locked during insertion and rotation. This locking strategy excludes other processes performing insertions from the entire subtree rooted by the parent of the critical node. Hence the number of concurrent insertions is quite limited. (Ellis has designed another parallel algorithm that requires fewer nodes to be locked and hence allows more insertions to take place concurrently [Ellis 1980b].)

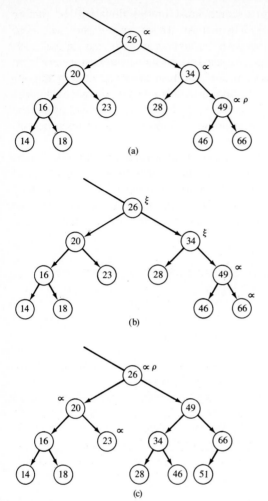

Figure 5-4 Parallel search and insertion in an AVL tree.

To summarize, Ellis's algorithm assigns to individual processes particular keys to be searched for or inserted. A number of searches and insertions can take place concurrently. The algorithm has three important weaknesses. First, it does not allow deletions to be performed. Second, it suffers from software lockout—processes performing insertions prevent other insertion processes from accessing entire subtrees. Third, the parallel algorithm has a lot of overhead. Even a search process has to lock and unlock every node it examines.

Manber and Ladner's Algorithm

Manber and Ladner [1982] have suggested a parallel searching algorithm for the tightly coupled multiprocessor model that allows deletions but sacrifices the AVL tree property. Manber and Ladner's algorithm does not try to maintain a balanced tree. They hope that the insertions and deletions will be random enough to maintain a reasonably balanced tree. (A study by Eppinger [1983] suggests that this hope is justified only under certain conditions.) In return for this concession, their algorithm is much simpler, because insertion and deletion processes do not have the responsibility of rotating the tree. An interesting feature of this algorithm is the use of maintenance processes to physically delete nodes. Manber and Ladner also show how to update the data field associated with a particular key. We describe their concurrent update operation, even though updating is not a dictionary operation.

Every node v in the search tree has 10 fields associated with it:

$Key(v)$, according to which the nodes are ordered
$Data(v)$, a pointer to the data associated with the key
$Left(v)$, a pointer to the left subtree of v
$Right(v)$, a pointer to the right subtree of v
$Parent(v)$, a pointer to the parent of v
$Garbage(v)$, set to *true* when the node has been removed from the tree
$Redundant(v)$, set to *true* when the node has a copy in the tree
$Copy(v)$, which when *true* means v is a copy of another node
$Userlock(v)$, a lock that can be set by a delete process
$Mlock(v)$, a lock that can be set by a maintenance process

The root of the tree is a special-purpose node with key ∞. It contains no data and cannot be deleted. This simplifies the parallel algorithms.

There are two reasons why nodes need to be locked. First, a node needs to be locked to ensure that only one process at a time updates its data. Second, nodes need to be locked to make sure that two or more processes cannot attempt to change the shape of the tree at the same place at the same time. These two purposes are not independent; for example, one process cannot be allowed to delete a node while another process is updating it. Thus, one kind of lock, the *userlock*, is used for both purposes.

Every key is associated with a unique node. If the key is in the tree, its associated node is the node containing the key. If the key is not in the tree, its associated node is the node that would be the parent of the key node if the key were to be inserted into the tree. Hence internal nodes are associated with a single key, while leaf nodes may be associated with a number of keys corresponding to some range of values. The basic operations of update, insert, and delete involve a single node. These operations begin by performing a **strong search** (to be defined later), which finds the node associated with the key and locks the *userlock* of that node. If the *userlock* is already locked, the process performing the basic operation is blocked until it is unlocked. Only one basic operation can be performed on a particular node at one time. In contrast to Ellis's algorithm, however, all basic operations can be performed by locking only one node. This provides much more latitude for concurrency. Note that except for the data field, all fields of a node may be examined while the node is locked.

This allows a search to work its way through a locked interior node.

Basic User Operations. The basic user operations are strong search, weak search, update, insert, and delete. This section describes these basic operations.

Since searching does not change the tree, its parallel implementation is quite similar to the sequential algorithm. However, there is one important factor to consider. What happens if the search is performed and the result of the search reported, but another process has modified the tree between the time when the search was terminated and the result was reported? Sometimes this factor is irrelevant to the computation; at other times this factor may be crucial. Hence we define two different search algorithms.

Weak search returns a result that is not guaranteed to be up to date. Weak search should be used whenever possible, because it does not need to lock any nodes. Thus the parallel algorithm has no overhead and a process performing a weak search interferes with no other processes.

Strong search searches for a given key and returns the node v associated with that key. To make sure that v will remain the node associated with the key as long as it is needed, the process performs a weak search and then locks v. After node v is locked, however, a check must be done to make sure that v is still the node associated with the key. Between the time when v is found and when v is locked, three different events could invalidate the association between the key and v. First, v might have been removed from the tree [that is, $garbage(v)$ set to $true$]. Second, if $key(v)$ is not the key being searched for, another node might have been inserted into the tree that is now the node associated with the key. Third, v might have become a redundant node [that is, $redundant(v)$ set to $true$]. Redundant nodes are created as a side effect of deletions. To summarize, when strong search returns a node, that node is guaranteed to be the node associated with the key and that node is locked. The node is unlocked only after the operation—update, insert, or delete—is completed.

In Manber and Ladner's algorithm deletion is simple, because all the deletion process has to do is logically delete the node. Maintenance processes take care of physically deleting the node. The delete process uses strong search to find the node v associated with the key and lock it. Assuming the key values match, $data(v)$ is set to nil. The delete process puts v on a special maintenance list, called the *delete list*, so that a maintenance process can come along later and physically delete the node.

Concurrent insertion begins by performing a strong search for the node v associated with the key. If a node associated with the key is found, it may have been logically deleted [that is, $data(v)$ is nil]. If so, the process adds the new data field, unlocks v, and terminates. If v has not been logically deleted, the process reports that the node already exists, unlocks v, and terminates. Assuming the key is not in the tree, then v is the leaf node which is to become the parent of the inserted node, and the insertion is performed identically to the sequential algorithm. After the correct child pointer of v has been set to point to the inserted node, the process unlocks v and terminates.

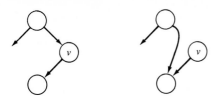

Figure 5-5 Deletion of a node with one child.

Concurrent update is straightforward. First the appropriate node v is found through strong search. Then the process updates the data field, unlocks v, and terminates.

Maintenance Processes. The maintenance processes physically delete nodes from the search tree. Recall that when a node is deleted, a pointer to it is put on a deletion list. Idle maintenance processes access this list to get pointers to nodes needing deletion.

If node v needs deletion and v has only one child, the deletion is simple. The pointer from v's parent to v is redirected to v's child, and v is effectively detached from the tree (see Figure 5-5). (If v is a leaf node, then the pointer from v's parent to v is made nil.)

How soon can node v be reused? Although v is no longer a part of the tree, another process may have been accessing v while it was being detached. The purpose of the boolean $garbage(v)$ is to alert such a process that v is no longer a part of the tree. Since $parent(v)$ still points to its parent, an otherwise stranded search process can get back into the tree and continue.

The previous paragraph indicates that v cannot be reused immediately. Manber and Ladner suggest that the garbage collection algorithm of Kung and Lehman [1980] be used. This algorithm uses three lists of nodes: the passive garbage list, the active garbage list, and the available list. When a maintenance process first removes a node from the tree, the node is put on the passive garbage list. This list grows until a maintenance process is ready to perform garbage collection. Garbage collection begins by copying a group of nodes from the passive garbage list into the active garbage list and noting which processes are active at the time of the copy. When all these processes have terminated, the active garbage list is appended to the available list. Since the only processes that could possibly access nodes on the passive garbage list are those that were active when the copy to the active list was made, once these processes have terminated, the nodes are truly inactive and suitable for reuse.

The physical deletion algorithm is more complicated if the node v to be deleted has two children. Conceptually the deletion is performed in two steps, as illustrated in Figure 5-6. First, the node w having the largest key less than $key(v)$ is found (Figure 5-6a). Node w has at most one child (since it cannot have a right child). Hence it is easy to delete w by using the algorithm described earlier (Figure 5-6b). The second step is to replace v with w (Figure 5-6c). Since w is the immediate predecessor of v, the tree remains consistent. However, we

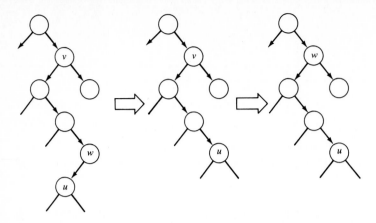

Figure 5-6 Conceptual idea of deletion of a node with two children.

must implement the deletion in a slightly different way to make sure that a copy of w is always accessible to a process that may be searching for it.

The algorithm used by Manber and Ladner is illustrated in Figure 5-7. First, create a copy of w, called w'. Note that $data(w') = data(w)$ and $copy(w') = true$. Second, set $left(w')$ to point to $left(v)$ and $right(w')$ to point to $right(v)$. Third, set $right(w)$ to point to w', and set $redundant(w)$ to $true$. What happens if a process searching for node w is on the way from node u to w when this operation takes place? This process may, for example, want to insert to the right of node w. When the process encounters node w, it will find that $redundant(w) = true$ and will therefore follow $right(w)$ to w', where it will continue to node t. Fourth, remove node v by setting the appropriate child pointer of $parent(v)$ to point to w' and setting $garbage(v)$ to $true$.

Node w cannot be deleted immediately, because there may be processes along the way from u to w that are looking for w. A method similar to the garbage collection algorithm already described solves the problem. We put node w on a *passive redundancy list*. Nodes on the passive redundancy become available for physical deletion only after all the processes alive at the time of the substitution terminate. Once w is available for physical deletion, the process is straightforward, since w has only a single child. The phases of the deletion process are summarized in Figure 5-8.

5-3 SUMMARY

It is hard to use processors efficiently if you want to speed up a single search operation. The sequential algorithm has logarithmic complexity, and the speedup achieved by the parallel algorithm can be logarithmic only in the number of processors used. Hence the goal should be to perform a series of searches, insertions, and deletions as quickly as possible.

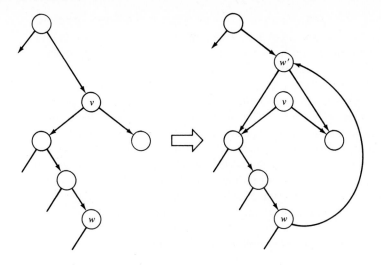

Figure 5-7 Deletion of a node with two children.

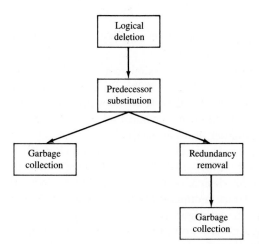

Figure 5-8 Phases of the deletion process of Manber and Ladner's algorithm.

This chapter has described parallel algorithms to implement searching, inserting, and deleting on tightly coupled multiprocessors. Both algorithms associate processes with particular operations to be performed. Locks must be used to keep processes from interfering with each other during critical operations. A primary difference between the two algorithms is that Ellis's algorithm may require locking a number of nodes logarithmic in the size of the tree, whereas Manber and Ladner's algorithm never requires more than a single node to be locked.

Ellis's algorithm allows searching and inserting to take place in AVL trees. A process inserting a key must lock the subtree rooted by the parent of the node to be rotated. Hence the amount of concurrency among inserting processes is limited. Search processes must lock and unlock every node they examine. Although more than one search process can share such a lock, the frequency of the lock-unlock operation adds to the overhead of the parallel algorithm.

Manber and Ladner's algorithm requires that no search, insert, delete, or update process lock more than a single node of the search tree. This greatly improves the potential efficiency of the parallel algorithm. The delete process simply marks a node for deletion. Separate maintenance processes, which have to lock three nodes, perform the physical deletion of nodes from the search tree. The algorithm does not maintain a balanced tree. Manber and Ladner hope that the insertions and deletions will be random enough to keep the tree reasonably balanced. Evidence published after Manber and Ladner's work suggests that this hope is justified only under certain conditions [Eppinger 1983].

Our description of parallel searching has avoided the topic of memory bank conflicts. This important issue is addressed in the exercises.

BIBLIOGRAPHIC NOTES

Most books on data structures, including Wirth [1976] and Helman and Veroff [1986], discuss AVL trees.

This chapter does not describe Ellis's second parallel search and insertion algorithm for AVL trees. It can be found in Ellis [1980b]. In addition to her work on AVL trees, Ellis [1980a] has described algorithms allowing concurrent search and insertion in 2-3 trees. Code for all Manber and Ladner's procedures described in this chapter appears in Manber and Ladner [1982].

Two early papers discussing concurrent insertion into and balancing of binary search trees are Wong and Chang [1974] and Chang [1974]. Work on concurrent access to trees holds interest for designers of data base systems. Kung and Lehman [1980] discuss the concurrent manipulation of binary search trees; Lehman and Yao [1981] describe concurrent algorithms for B* trees. These algorithms have the desirable property that tree rotations require no more than a small constant number of nodes to be locked.

Baer, Du, and Ladner [1983] investigate a number of algorithms to perform batch searching on processor arrays and tightly coupled multiprocessors. Their paper focuses on the important problem of avoiding memory bank conflicts. Carey and Thompson [1984] have proposed a systolic algorithm for implementing search trees on MIMD computers. Ottman, Rosenberg, and Stockmeyer [1982] have designed searching algorithms suitable for implementation in VLSI. Potter [1985] briefly describes the ease of searching on Goodyear's Massively Parallel Processor, an SIMD-MC^2 computer. Ramamoorthy, Turner, and Wah [1978] have proposed a searching machine based upon associative memory.

EXERCISES

5-1 Fill in the inductive argument of the second step of the proof of Theorem 5-1 by proving that after k parallel computation steps there must be one or more contiguous unexamined segments of the table with length at least

$$\frac{n+1}{(p+1)^k} - 1$$

*5-2 Given an $n \times n$ SIMD-MC^2 model containing a sorted list of n^2 items and $n \log n$ items to search for, what is the time needed to complete all $n \log n$ searches?

5-3 Is Ellis's algorithm pipelined, partitioned, or relaxed? Explain your answer.

5-4 In the context of Ellis's algorithm, provide an example showing how two processes performing rotations could contend with each other if the entire subtree rooted by the parent of the node to be rotated were not locked.

5-5 Is Manber and Ladner's algorithm pipelined, partitioned, or relaxed? Explain your answer.

5-6 Provide an example showing why physical deletion of a node with two children cannot be performed as shown in Figure 5-6.

5-7 Provide an example showing why deleted nodes must be put on the passive garbage list and the active garbage list before being put on the list of available nodes.

5-8 In the context of Manber and Ladner's algorithm, explain which nodes need to be locked by a maintenance process and why.

5-9 Assume a tightly coupled multiprocessor with four CPUs and four memory banks. Assume a table A of n keys is to be searched, where $a[1] < a[2] < \cdots < a[n]$, and $a[i]$ is stored in memory bank i modulo 4. Furthermore, make the simplifying assumption that the processors work in lockstep and that every processor fetches and compares an element of A every memory cycle. If two processors P_i and P_j, where $i < j$, try to access the same memory bank in the same cycle, processor P_i gets the value it desires, while P_j must repeat the access attempt on the next cycle. It is helpful to think of the binary search of an array as the traversal of a binary tree. See Figure 5-9 for an example of a binary tree representing an array of 20 keys. Given all these assumptions, suppose that $n = 20$ and $a[i] = i$ for $1 \le i \le 20$. Suppose processor P_0 is searching for value 13, P_1 is searching for 4, P_2 is searching for 10, and P_3 is searching for 17. Draw a table showing the execution of the batch search for these keys. In particular, indicate which memory bank a processor accesses (or fails to access) each memory cycle.

5-10 Repeat the previous exercise, with the following modifications. Suppose $n = 31$ and $a[i] = i$ for $1 \le i \le 31$.

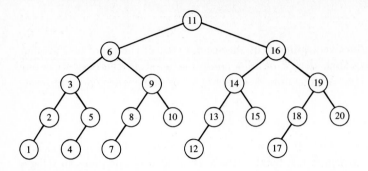

Figure 5-9 A binary search of a 20-key table can be viewed as a search of this tree.

MATRIX MULTIPLICATION

Like sorting and searching, matrix multiplication is a fundamental component of many numerical and nonnumerical algorithms. This chapter examines several parallel algorithms to perform matrix multiplication. Section 6-1 begins with Gentleman's proof that matrix multiplication on the SIMD-MC^2 model has complexity $\Omega(n)$ and continues with the description of an optimal matrix multiplication algorithm for this model. Matrix multiplication can be done in logarithmic time on the SIMD-CC and SIMD-PS models, however, and the section concludes with a presentation of Dekel, Nassimi, and Sahni's $\Theta(\log n)$ matrix multiplication algorithm for the SIMD-CC model. Section 6-2 discusses matrix multiplication on multiprocessors. There are three straightforward ways to parallelize the sequential algorithm on this model. The grain size lemma is used to help determine which way leads to the best speedup. None of the straightforward algorithms is suitable for the loosely coupled multiprocessor model, since they require processors to fetch too many nonlocal operands. Another parallel algorithm is developed for this model.

Sequential Matrix Multiplication

The product of an $l \times m$ matrix $\mathbf{A}$ and an $m \times n$ matrix $\mathbf{B}$ is an $l \times n$ matrix $\mathbf{C}$ whose elements are defined by

$$c_{i,j} = \sum_{k=0}^{m-1} a_{i,k} b_{k,j}$$

A sequential algorithm implementing matrix multiplication follows.

123

MATRIX MULTIPLICATION (SISD):
```
begin
    for i ← 0 to l − 1 do
        for j ← 0 to n − 1 do
            t ← 0
            for k ← 0 to m − 1 do
                t ← t + a_{i,k} × a_{k,j}
            endfor
            c_{i,j} ← t
        endfor
    endfor
end
```

The algorithm requires $l \times m \times n$ additions and multiplications. The time complexity of multiplying two $n \times n$ matrices using this sequential algorithm is clearly $\Theta(n^3)$. Sequential matrix multiplication algorithms with a lower time complexity have been developed, such as Strassen's algorithm, but every algorithm developed in this chapter is a parallelization of the straightforward algorithm.

6-1 ALGORITHMS FOR PROCESSOR ARRAYS

Matrix Multiplication on the SIMD-MC² Model

A Lower Bound. Gentleman [1978] has shown that multiplication of two $n \times n$ matrices on the SIMD-MC² model requires $\Omega(n)$ data routing steps.

Definition 6-1 Given a data item originally available at a single processor in some model of parallel computation, let the function $\sigma(k)$ be the maximum number of processors to which the data can be transmitted in k or fewer data routing steps.

For example, in the SIMD-MC² model $\sigma(0) = 1$, $\sigma(1) = 5$, $\sigma(2) = 13$, and in general $\sigma(k) = 2k^2 + 2k + 1$.

Lemma 6-1 (See Gentleman [1978].) Suppose that two $n \times n$ matrices **A** and **B** are to be multiplied, and assume that every element of **A** and **B** is stored exactly once and that no processing element contains more than one element of either matrix. If we ignore any data broadcasting facility, multiplying **A** and **B** to produce the $n \times n$ matrix **C** requires at least s data routing steps, where $\sigma(2s) \geq n^2$.

Proof. Consider an arbitrary element $c_{i,j}$ of the product matrix. This element is the inner product of row i of matrix **A** and column j of matrix **B**. There must be a path from the processors where each of these elements is stored to the processor where the result $c_{i,j}$ is stored. Let s denote the length of the longest such path (Figure 6-1a). In other words, the creation of matrix product $\mathbf{C}$ takes at least s data routing steps.

Note that these paths also can be used to define a set of paths of length at most $2s$ from any element $b_{u,v}$ to every element $a_{i,j}$, where $1 \leq i, j \leq n$. This is because there is a path of length at most s from $b_{u,v}$ to the processor where $c_{i,v}$ is found, and there is also a path of length at most s from $a_{i,j}$ to $c_{i,v}$ (Figure 6-1b). Hence there is a path of length at most $2s$ from any element $b_{u,v}$ to every element $a_{i,j}$. Similarly these paths define a set of paths of length at most $2s$ from any element $a_{u,v}$ to every element $b_{i,j}$, where $1 \leq i, j \leq n$.

The n^2 elements of $\mathbf{A}$ are stored in unique processors. Since there exist paths of length at most $2s$ from the processor storing $b_{u,v}$ to the processors storing the elements of $\mathbf{A}$, it follows from the definition of σ that $\sigma(2s) \geq n^2$. ∎

Theorem 6-1 (See Gentleman [1978].) Matrix multiplication on the SIMD-MC^2 model requires $\Omega(n)$, or for large values of n, approximately $s \geq 0.35n$, data routing steps.

Proof. From the previous lemma we have $\sigma(2s) \geq n^2$. On the SIMD-MC^2 model $\sigma(s) = 2s^2 + 2s + 1$. Hence $\sigma(2s) = 2(2s)^2 + 2(2s) + 1$. Combining the two yields

$$8s^2 + 4s + 1 \geq n^2$$
$$\Rightarrow \qquad s^2 + \frac{s}{2} + \frac{1}{8} \geq \frac{n^2}{8}$$
$$\Rightarrow \qquad \left(s + \frac{1}{4}\right)^2 + \frac{1}{16} \geq \frac{n^2}{8}$$
$$\Rightarrow \qquad s \geq \frac{\sqrt{2n^2 - 1}}{4} - \frac{1}{4} \qquad ∎$$

An Optimal Algorithm. Given an SIMD-MC^2 model with wraparound connections, it is relatively easy to devise an algorithm that uses n^2 processors to multiply two $n \times n$ arrays in $\Theta(n)$ time.

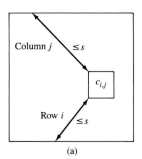
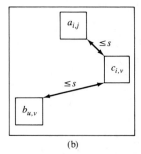

Figure 6-1 Gentleman's proof. (a) No path to $c_{i,j}$ has length more than s. (b) Element $b_{u,v}$ cannot be more than $2s$ data routings from element $a_{i,j}$.

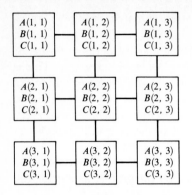

Figure 6-2 Every processing element $P(i,j)$ contains local scalars $A(i,j)$, $B(i,j)$, and $C(i,j)$.

The key insight is to realize that since n^3 multiplications are required (assuming the straightforward algorithm), the only way that n^2 processing elements can complete the multiplication in $\Theta(n)$ time is for $\Theta(n^2)$ processing elements to be contributing toward the result at every step. Examine the initial allocation of matrix elements to processing elements, illustrated in Figure 6-2. The processing element located at row i, column j in the mesh contains $a_{i,j}$ and $b_{i,j}$. Note that in this original state only n processing elements contain a pair that needs to be multiplied. However, it is possible to stagger matrices **A** and **B** so that at every step of the algorithm an upward rotation of the elements in **B** and a leftward rotation of the elements of **A** presents each processing element with a pair of values to be multiplied. Processing element $P(i,j)$ performs the additions that will form the dot product $c_{i,j}$. This insight is the basis of the parallel algorithm that follows.

When the algorithm begins execution, $A(i,j) = a_{i,j}$ and $B(i,j) = b_{i,j}$. The first phase of the parallel algorithm staggers the two matrices. The second phase computes all the products $a_{i,k} \times b_{k,j}$ and accumulates the sums. When phase 2 has completed, $C(i,j) = \Sigma_{k=1}^{n} A(i,k) \times B(k,j)$. The parallel algorithm follows.

```
MATRIX MULTIPLICATION (SIMD-MC²):
{A(i,j) contains a_{i,j} and B(i,j) contains b_{i,j}}
begin
   {Stagger matrices}
   for k ← 1 to n − 1 do
      for all P(i,j) do
         if i > k then
            A(i,j) ⇐ A(i, j + 1)
         endif
         if j > k then
            B(i,j) ⇐ B(i + 1, j)
         endif
      endfor
   endfor
   {Compute dot products}
   for all P(i,j) do
      C(i,j) ← A(i,j) × B(i,j)
   endfor
```

$P(0)$	$A(0) = a_{0,0}$	$B(0) = b_{0,0}$
$P(1)$	$A(1) = a_{0,1}$	$B(1) = b_{0,1}$
$P(2)$	$A(2) = a_{1,0}$	$B(2) = b_{1,0}$
$P(3)$	$A(3) = a_{1,1}$	$B(3) = b_{1,1}$
$P(4)$		
$P(5)$		
$P(6)$		
$P(7)$		

Figure 6-3 Initial allocation of matrix elements to processing elements on SIMD-CC model, where $n = 2$, $q = 1$, and $p = 8$.

```
for k ← 1 to n − 1 do
    for all P(i, j)
        A(i, j) ⇐ A(i, j + 1)
        B(i, j) ⇐ B(i + 1, j)
        C(i, j) ← C(i, j) + A(i, j) × B(i, j)
    endfor
endfor
end
```

Matrix Multiplication on the SIMD-CC Model

Theorem 6-2 (See Dekel, Nassimi, and Sahni [1981].) Given the SIMD-CC model with $n^3 = 2^{3q}$ processors, two $n \times n$ matrices can be multiplied in $\Theta(\log n)$ time.

The key to the algorithm of Dekel, Nassimi, and Sahni is the data routing strategy; $5q = 5 \log n$ routing steps are sufficient to broadcast the initial values through the processor array and to combine the results.

The processing elements should be thought of as filling an $n \times n \times n$ lattice. Processor $P(x)$, where $0 \le x \le 2^{3q} - 1$, has local memory locations $A(x)$, $B(x)$, $C(x)$, $S(x)$, and $T(x)$.

When the parallel algorithm begins execution, matrix elements $a_{i,j}$ and $b_{i,j}$, for $0 \le i, j \le n - 1$, are stored in $A(2^q i + j)$ and $B(2^q i + j)$ (Figure 6-3). After the parallel algorithm has completed, matrix elements $c_{i,j}$, for $0 \le i, j \le n - 1$, should be stored in $C(2^q i + j)$.

The algorithm has three distinct phases. During the first phase the $a_{i,j}$ and $b_{i,j}$ must be broadcast to the rest of the processors. After this **for** loop,

$$\left. \begin{array}{l} A(2^{2q} k + 2^q i + j) = a_{i,j} \\ B(2^{2q} k + 2^q i + j) = b_{i,j} \end{array} \right\} \text{ for } 0 \le k \le n - 1$$

After the second **for** loop,

$$A(2^{2q} k + 2^q i + j) = a_{i,k} \quad \text{for } 0 \le j \le n - 1$$

$BIT(9,0) = 1$	$BIT\_COMPLEMENT(9,0) = 8$
$BIT(9,1) = 0$	$BIT\_COMPLEMENT(9,1) = 11$
$BIT(9,2) = 0$	$BIT\_COMPLEMENT(9,2) = 13$
$BIT(9,3) = 1$	$BIT\_COMPLEMENT(9,3) = 1$
$BIT(9,4) = 0$	$BIT\_COMPLEMENT(9,4) = 25$
$BIT(9,5) = 0$	$BIT\_COMPLEMENT(9,5) = 41$

Figure 6-4 Functions BIT and BIT_COMPLEMENT.

After the third for loop,

$$B(2^{2q}k + 2^q i + j) = b_{k,j} \quad \text{for } 0 \le i \le n - 1.$$

There are n^3 multiplications to be performed and n^3 processing elements available. Because of the broadcasting done in the first phase, all multiplications $a_{i,k}b_{k,j}$ can be done simultaneously during phase 2. The third phase of the algorithm routes and sums the products.

Two new functions, **BIT** and **BIT_COMPLEMENT**, are used by this algorithm. Function **BIT**, passed integer arguments m and l, returns the value of the lth bit in the binary representation of m. Function **BIT_COMPLEMENT**, passed integer arguments m and l, returns the value of integer formed by complementing the value of bit l in the binary representation of m. These functions are illustrated in Figure 6-4.

The parallel algorithm follows.

```
MATRIX MULTIPLICATION (SIMD-CC):
begin
  {Phase 1: Broadcast A and B}
  for l ← 3q − 1 downto 2q do
    for all P(m), where BIT(m, l) = 1 do
      T(m) ← BIT_COMPLEMENT(m, l)
      A(m) ⇐ A(T(m))
      B(m) ⇐ B(T(m))
    endfor
  endfor
  for l ← q − 1 downto 0 do
    for all P(m), where BIT(m, l) ≠ BIT(m, 2q + l) do
      T(m) ← BIT_COMPLEMENT(m, l)
      A(m) ⇐ A(T(m))
    endfor
  endfor
  for l ← 2q − 1 downto q do
    for all P(m), where BIT(m, l) ≠ BIT(m, q + l) do
      T(m) ← BIT_COMPLEMENT(m, l)
      B(m) ⇐ B(T(m))
    endfor
  endfor
```

```
{Phase 2: Do the multiplications in parallel}
for all P(m) do
    C(m) ← A(m) × B(m)
endfor

{Phase 3: Sum the products}
for l ← 2q to 3q − 1 do
    for all P(m) do
        S(m) ← BIT_COMPLEMENT(m, l)
        T(m) ⇐ C(S(m))
        C(m) ← C(m) + T(m)
    endfor
endfor
end
```

Figure 6-5 illustrates the operation of this algorithm as it multiplies two 2×2 matrices on an eight-processor SIMD-CC model.

The first for loop requires $2q$ data routing steps. The last three for loops require q data routing steps each. Hence a total of $5q$ data routing steps are sufficient to multiply two matrices on the SIMD-CC model. The algorithm also uses one multiplication step and q addition steps. Clearly, then, the complexity of matrix multiplication on the SIMD-CC model is $\Theta(q) = \Theta(\log n)$, given n^3 processing elements.

Matrix Multiplication on the SIMD-PS Model

Theorem 6-3 (See Dekel, Nassimi, and Sahni [1981].) Given $n^3 = 2^{3q}$ processors on the SIMD-PS model, two $n \times n$ matrices can be multiplied in $\Theta(\log n)$ time.

The algorithm for the perfect shuffle model simulates the routes followed in the algorithm presented for the SIMD-CC model. In $13 \log n$ routing steps the SIMD-PS model can perform the same routings done by the SIMD-CC model in $5 \log n$ steps. Hence matrix multiplication on the SIMD-PS model has complexity $\Theta(\log n)$, given $n^3 = 2^{3q}$ processing elements.

6-2 ALGORITHMS FOR MULTIPROCESSORS

Matrix multiplication presents all sorts of opportunities for parallelization on multiprocessors. In the sequential algorithm at the beginning of the chapter, three for loops could be parallelized. That brings up an interesting question: When there are a number of nested loops, all suitable for parallelization, which one should be made parallel? The following lemma gives guidance in designing parallel algorithms for tightly coupled multiprocessors.

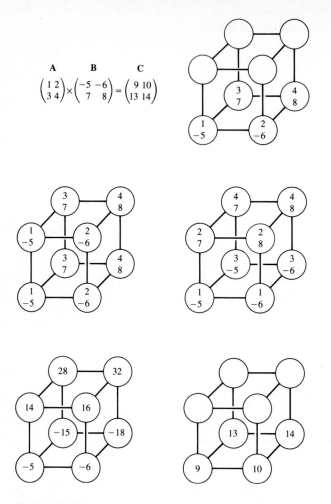

A   B   C

$$\begin{pmatrix} 1 & 2 \\ 3 & 4 \end{pmatrix} \times \begin{pmatrix} -5 & -6 \\ 7 & 8 \end{pmatrix} = \begin{pmatrix} 9 & 10 \\ 13 & 14 \end{pmatrix}$$

Figure 6-5 Matrix multiplication on the SIMD-CC model. (Quinn and Deo [1984]. Copyright © 1986 Association for Computing Machinery. Reprinted by permission.)

Lemma 6-2 (Grain Size Lemma) Given a tightly coupled multiprocessor with p active processors that must be synchronized via a single global semaphore, and given a task with worst-case time complexity $t(n)$ that requires synchronization after its completion, then the maximum possible speedup is $t(n)/\left\lceil \sqrt{2t(n) + \frac{1}{4}} + \frac{1}{2} \right\rceil$, or $O(\sqrt{t(n)})$.

Proof. To synchronize via a single global semaphore, each process must lock, increment, and unlock the semaphore. Without loss of generality, assume this takes 1 unit of time. Then the synchronization of p processors takes at least p units of time. The amount of processing that can take place during synchronization is, therefore, at least $p(p-1)/2$ time units of work (see Figure 6-6). When

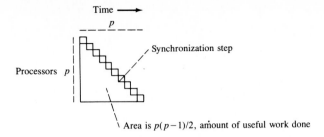

Area is $p(p-1)/2$, amount of useful work done

Figure 6-6 Processing during synchronization on the MIMD-TC model.

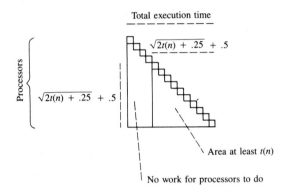

Figure 6-7 Adding processors may increase execution time.

$p = \left\lceil \sqrt{2t(n) + \frac{1}{4}} + \frac{1}{2} \right\rceil$, the amount of work that can be done is at least $t(n)$. Thus, $O(\sqrt{t(n)})$ processors are sufficient to minimize execution time. Adding more processors only adds to the total execution time (see Figure 6-7).

Alternate Proof. A task having worst-case time complexity $t(n)$ on a sequential computer will have worst-case complexity $\Omega(t(n)/p)$ on a parallel computer with p processors. To synchronize via a single global semaphore, each processor must have exclusive access to that semaphore. Hence synchronizing p processors via a single semaphore has time complexity $\Omega(p)$. Therefore, the complexity of any parallel algorithm solving the task must be $\Omega(\max\{t(n)/p, p\})$. This function is minimized when $p = \sqrt{t(n)}$. Since the complexity of the parallel algorithm, then, is $\Omega(\sqrt{t(n)})$, the maximum possible speedup is $O(\sqrt{t(n)})$. ∎

What are the practical consequences of this lemma regarding parallel matrix multiplication? The innermost for loop has complexity $\Theta(n)$. If this loop is parallelized, the maximum speedup achievable is $O(\sqrt{n})$. The middle for loop has complexity $\Theta(n^2)$. If this loop is made parallel, the algorithm may achieve $O(n)$ speedup. Finally, the outer for loop has complexity $\Theta(n^3)$. Making this loop parallel increases the potential speedup to $O(n^{1.5})$. Of course, other factors, such as partitionability of the algorithm and contention for elements of **A** and

B, may make this much speedup impossible to achieve, but in general the grain size lemma tells us that when given a choice, always parallelize the outermost loop.

A parallel matrix multiplication algorithm for the tightly coupled multiprocessor model follows. The variables $i(m)$, $j(m)$, $k(m)$, and $t(m)$ represent scalars local to process $P(m)$. Parentheses have been used instead of subscripts to make the algorithm more legible.

```
MATRIX MULTIPLICATION (TIGHTLY COUPLED MULTIPROCESSOR):
begin
    for all P(m), where 1 ≤ m ≤ p do
        for i(m) ← m to n step p do
            for j(m) ← i to n do
                t(m) ← 0
                for k(m) ← 1 to n do
                    t(m) ← t(m) + a_{i(m),k(m)} × b_{k(m),j(m)}
                endfor
                c_{i(m),j(m)} ← t(m)
            endfor
        endfor
    endfor
end
```

What is the complexity of this algorithm? Each process calculates n/p rows of matrix **C**; the time needed to calculate a single row is $\Theta(n^2)$. The processes synchronize exactly once; synchronization overhead, then, is $\Theta(p)$. Hence the complexity of this parallel algorithm is $\Theta(n^3/p + p)$. Note that since there are only n rows, at most n processes may execute this algorithm. If we ignore memory contention, we can expect speedup to be nearly linear. Figure 6-8 plots the speedup achieved by a single-PEM Denelcor HEP executing a HEP FORTRAN 77 version of this parallel algorithm under the HEP/UPX operating system on a 36×36 matrix.

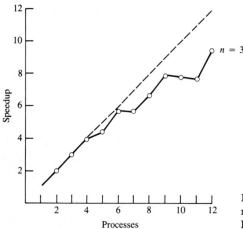

Figure 6-8 Speedup of parallel matrix multiplication algorithm on single-PEM Denelcor HEP.

Is it realistic to ignore memory access times? This assumption is safe on tightly coupled multiprocessors, where every global memory cell is an equal distance from every processor, but it is dangerous on loosely coupled multiprocessors, where some matrix elements may be much easier to access than others. Recall that on loosely coupled multiprocessors, such as Cm*, it is important to keep as many memory references as possible local. The above algorithm does not do a very good job of this. Indeed, not only must a typical process access n/p rows of **A**, but also it must access every element of **B** n/p times! Only a single addition and a single multiplication occur for every element of **B** fetched. This is not a very good ratio. Implementation of the above algorithm on Cm* would yield a very small speedup.

Another way must be found to partition the problem. One attractive method is to take advantage of block matrix multiplication. Assume that **A** and **B** are both $n \times n$ matrices, where $n = 2k$. Then **A** and **B** can be thought of as conglomerates of four smaller matrices, each of size $k \times k$:

$$\mathbf{A} = \begin{pmatrix} A_{11} & A_{12} \\ A_{21} & A_{22} \end{pmatrix} \quad \mathbf{B} = \begin{pmatrix} B_{11} & B_{12} \\ B_{21} & B_{22} \end{pmatrix}$$

Given this partitioning of **A** and **B** into blocks, **C** is defined as follows:

$$\mathbf{C} = \begin{pmatrix} C_{11} & C_{12} \\ C_{21} & C_{22} \end{pmatrix} = \begin{pmatrix} A_{11}B_{11} + A_{12}B_{21} & A_{11}B_{12} + A_{12}B_{22} \\ A_{21}B_{11} + A_{22}B_{21} & A_{21}B_{12} + A_{22}B_{22} \end{pmatrix}$$

If we assign processes to do the block matrix multiplications, then the number of multiplications and additions per matrix element fetch increases. For example, assume that there are $p = (n/k)^2$ processes. Then the matrix multiplication is performed by dividing **A** and **B** into p blocks of size $k \times k$. Each block multiplication requires $2k^2$ memory fetches, k^3 additions, and k^3 multiplications. The number of arithmetic operations per memory access has risen from 2, in the previous algorithm, to $k = n/\sqrt{p}$, in the new algorithm, a significant improvement. An example of this block matrix approach to parallel matrix multiplication is shown in Figure 6-9.

Ostlund, Hibbard, and Whiteside [1982] have implemented this matrix multiplication algorithm on Cm* for various matrix sizes. The results of their experiment are shown in Figure 6-10.

6-3 SUMMARY

A number of interesting parallel algorithms for matrix multiplication have been developed. Gentleman has shown that multiplying two $n \times n$ matrices on the SIMD-MC^2 model has complexity $\Omega(n)$. Dekel, Nassimi, and Sahni have devised efficient routing algorithms allowing $n^3 = 2^{3q}$ processing elements on either an SIMD-CC or an SIMD-PS model to multiply two $n \times n$ matrices in $\Theta(\log n)$ time. Matrix multiplication provides all sorts of opportunities for parallelization on multiprocessors. The grain size lemma can be used to help sort out the

$$
\begin{matrix}
\text{A} & \text{B} & \text{C}
\end{matrix}
$$

$$
\begin{pmatrix}
1 & 0 & 2 & 3 \\
4 & -1 & 1 & 5 \\
-2 & -3 & -4 & 2 \\
-1 & 2 & 0 & 0
\end{pmatrix}
\begin{pmatrix}
-1 & 1 & 2 & -3 \\
-5 & -4 & 2 & -2 \\
3 & -1 & 0 & 2 \\
1 & 0 & 4 & 5
\end{pmatrix}
=
\begin{pmatrix}
8 & -1 & 14 & 16 \\
9 & 7 & 26 & 17 \\
7 & 14 & -2 & 14 \\
-9 & -9 & 2 & -1
\end{pmatrix}
$$

STEP 1: Compute $C_{i,j} = A_{i,1}B_{1,j}$

$$
\begin{pmatrix}
1 & 0 \\
4 & -1
\end{pmatrix}
\begin{pmatrix}
-1 & 1 \\
-5 & -4
\end{pmatrix}
\Bigg|
\begin{pmatrix}
1 & 0 \\
4 & -1
\end{pmatrix}
\begin{pmatrix}
2 & -3 \\
2 & -2
\end{pmatrix}
=
\begin{pmatrix}
-1 & 1 & | & 2 & -3 \\
1 & 8 & | & 6 & -10 \\
\hline
17 & 10 & | & -10 & 12 \\
0 & -9 & | & 2 & -1
\end{pmatrix}
$$

$$
\begin{pmatrix}
-2 & -3 \\
-1 & 2
\end{pmatrix}
\begin{pmatrix}
-1 & 1 \\
-5 & -4
\end{pmatrix}
\Bigg|
\begin{pmatrix}
-2 & -3 \\
-1 & 2
\end{pmatrix}
\begin{pmatrix}
2 & -3 \\
2 & -2
\end{pmatrix}
$$

STEP 2: Compute $C_{i,j} + A_{i,2}B_{2,j}$

$$
\begin{pmatrix}
2 & 3 \\
1 & 5
\end{pmatrix}
\begin{pmatrix}
3 & -1 \\
1 & 0
\end{pmatrix}
\Bigg|
\begin{pmatrix}
2 & 3 \\
1 & 5
\end{pmatrix}
\begin{pmatrix}
0 & 2 \\
4 & 5
\end{pmatrix}
=
\begin{pmatrix}
8 & -1 & | & 14 & 16 \\
9 & 7 & | & 26 & 17 \\
\hline
7 & 14 & | & -2 & 14 \\
-9 & -9 & | & 2 & -1
\end{pmatrix}
$$

$$
\begin{pmatrix}
-4 & 2 \\
0 & 0
\end{pmatrix}
\begin{pmatrix}
3 & -1 \\
1 & 0
\end{pmatrix}
\Bigg|
\begin{pmatrix}
-4 & 2 \\
0 & 0
\end{pmatrix}
\begin{pmatrix}
0 & 2 \\
4 & 5
\end{pmatrix}
$$

Figure 6-9 Block matrix approach to parallel matrix multiplication.

good opportunities from the bad. Developing an efficient matrix multiplication algorithm for the loosely coupled multiprocessor model is complicated by the nonhomogeneous memory structure. Reshaping the algorithm in terms of block matrix multiplication increases the ratio of arithmetic operations to nonlocal memory accesses and leads to a more efficient algorithm.

BIBLIOGRAPHIC NOTES

All the matrix multiplication algorithms presented in this chapter are paralleliza-tions of the most common sequential algorithm. A parallelization of Strassen's algorithm appears in Chandra [1976].

Ramakrishnan and Varman [1984] have designed an optimal algorithm for performing matrix multiplication on the SIMD-MC[1] model.

Hwang and Cheng [1982] have designed a VLSI block matrix multiplication algorithm that illustrates how matrices can be multiplied when there are far fewer

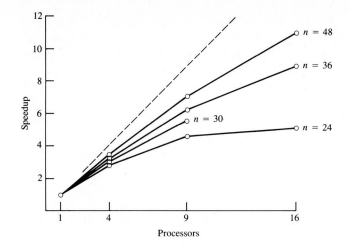

Figure 6-10 Speedup of matrix multiplication algorithm on Cm*.

processing elements than matrix elements. VLSI algorithms to perform matrix multiplication or related algorithms have also been published by Horowitz [1979], Kulkarni and Yen [1982], Leiserson [1983], and Ullman [1984].

EXERCISES

6-1 Write an $\Theta(\log n)$ matrix multiplication algorithm for the SIMD-SM-R model. Assume that $n = 2^k$, where k is a positive integer.

6-2 Illustrate $\sigma(0)$, $\sigma(1)$, and $\sigma(2)$ for the SIMD-MC^2 model.

6-3 Prove that for the SIMD-MC^2 model $\sigma(k) = 2k^2 + 2k + 1$.

*6-4 What is $\sigma(k)$ for the SIMD-CC model?

6-5 Gentleman's theorem assumes that every element of **A** and **B** is stored exactly once in the parallel computer and that no processing element contains more than one element of either matrix. Are these assumptions realistic? Explain.

6-6 Add a postamble to the SIMD-MC^2 matrix multiplication algorithm that "unstaggers" matrices **A** and **B** so that at the end of the algorithm $A(i,j) = a_{i,j}$, $B(i,j) = b_{i,j}$, and $C(i,j) = c_{i,j}$.

6-7 Write a parallel algorithm that multiplies two $n \times n$ matrices in time $\Theta(n)$ on the SIMD-MC^2 model with no wraparound connections. Processing element $P(i,j)$ has local variables $A(i,j)$, $B(i,j)$, and $C(i,j)$. When the algorithm begins execution, $A(i,j) = a_{i,j}$ and $B(i,j) = b_{i,j}$. After $\Theta(n)$ steps, $C(i,j) = c_{i,j}$. No processing element should use more than a constant number of variables.

6-8 Write a parallel algorithm that transposes an $n \times n$ matrix in time $\Theta(n)$ on the SIMD-MC^2 model with wraparound connections. Processing element $P(i,j)$ has local variables $A(i,j)$ and $T(i,j)$, where $1 \leq i,j \leq n$. When the algorithm begins execution, $A(i,j) = a_{i,j}$. After $\Theta(n)$ steps, $T(i,j) = a_{j,i}$. No processing element should use more than a constant number of variables.

6-9 Repeat the previous exercise with all assumptions identical, except that the model does not have wraparound connections.

6-10 Determine the processor efficiency of the SIMD-CC matrix multiplication algorithm as a function of the matrix dimension n.

6-11 If $p \leq n$ processors are available for matrix multiplication on a tightly coupled multiprocessor, what would be the complexity of the algorithm resulting from parallelizing the middle for loop of the sequential algorithm? Would you expect this algorithm to achieve better or worse speedup than the algorithm presented in the book? Why?

6-12 Explain the irregularities in the speedup curve of the parallel matrix multiplication algorithm implemented on the HEP.

6-13 Design a matrix multiplication algorithm for a multicomputer with a hypercubic processor organization.

SEVEN

NUMERICAL ALGORITHMS

This chapter describes parallel algorithms to solve recurrence relations, partial differential equations, and systems of linear equations. A recurrence relation is an equation that expresses the value of a function at one point in terms of its value at other points. Hockney and Jesshope [1981] provide a number of instances in which recurrence relations occur in numerical analysis. These instances include the solution of linear equations through gaussian elimination, solutions of ordinary differential equations in time, and marching methods for the solution of differential equations in space.

Partial differential equations appear frequently in engineering, physics, chemistry, and other physical sciences. The case study considered is the simple, intuitive problem of finding the steady state temperature distribution of a thin rectangular metal plate with fixed boundary conditions. The SIMD-MC^2 model is ideally suited to solve this kind of problem. Efforts to solve partial differential equations on the loosely coupled multiprocessor Cm* are also described later in the chapter.

The third problem to be considered is the solution of systems of linear equations through the well-known gaussian elimination algorithm. Focusing on the first, most computationally intensive, stage of the algorithm, we show that a tightly coupled multiprocessor can efficiently perform this process, even with a relatively large number of processors.

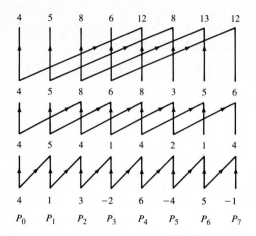

4   5   8   6   12   8   13   12

4   5   8   6   8   3   5   6

4   5   4   1   4   2   1   4

| 4 | 1 | 3 | −2 | 6 | −4 | 5 | −1 |
| P_0 | P_1 | P_2 | P_3 | P_4 | P_5 | P_6 | P_7 |

Figure 7-1 Parallel partial cascade sum on SIMD-CC model with eight processors.

7-1 SOLVING RECURRENCE RELATIONS ON PROCESSOR ARRAYS

A general first-order linear recurrence has the form

$$x_j = a_j x_{j-1} + d_j, \quad \text{for } j = 1, \, 2, \, \ldots, \, n$$

where the values $x_0, a_1, a_2, \ldots, a_n$ and $d_1, \ldots, d_n$ are given. It can be assumed, without loss of generality, that $x_0 = a_1 = 0$, since if these values are not zero, d_1 can be redefined to equal the sum $d_1 + a_1 x_0$. We make this assumption because it makes the analysis and the subsequent algorithm easier to understand. The primary source of information for this section is Hockney and Jesshope [1981].

Given a sequence of values $d_1, d_2, \ldots, d_n$, the **partial sum** x_i is defined to be $\Sigma_{j=1}^{i} d_j$. Finding all the partial sums $x_1, \ldots, x_n$ of a sequence of values is actually a special case of a first-order linear recurrence, the case in which all the a_j's are 1.

An SIMD-CC algorithm to compute the partial sums of a sequence of values appears below. The algorithm is commonly called the **partial cascade sum** method. Assume that the $n = 2^k$ values $d_1, d_2, \ldots, d_n$ are stored in processing elements $P_0, P_1, \ldots, P_{n-1}$. In other words, variable $d(0)$ in processing element P_0 contains d_1, and so on. When the algorithm completes, $x(i)$ will contain the partial sum x_{i+1}. An example of how this algorithm works appears in Figure 7-1.

```
PARTIAL CASCADE SUM (SIMD-CC):
begin
    for all P_i, where 0 ≤ i ≤ n − 1 do
        x(i) ← d(i)
    endfor
```

```
for i ← 0 to log n − 1 do
    for all Pⱼ, where 2ⁱ + 1 ≤ j ≤ n − 1 do
        t(j) ⇐ x(j − 2ⁱ)
        x(j) ← x(j) + t(j)
    endfor
endfor
end
```

Clearly the complexity of this algorithm is $\Theta(\log n)$ with n processing elements on the SIMD-CC model.

A quite similar algorithm can be developed to solve a general first-order linear recurrence relation. The algorithm is called **cyclic reduction**, and it begins by reducing relations between adjacent terms in the sequence into relations between every other term in the sequence. In the second step relations between terms two apart in the sequence are reduced to relations between terms four apart in the sequence. After a logarithmic number of reduction steps, the terms have taken on their final value, because they are related only to constant values and not to other terms.

Consider two successive terms of the recurrence relation:

$$x_j = a_j x_{j-1} + d_j \qquad (7.1)$$

$$x_{j-1} = a_{j-1} x_{j-2} + d_{j-1} \qquad (7.2)$$

If we solve equation (7.1) in terms of equation (7.2), we get

$$x_j = a_j a_{j-1} x_{j-2} + a_j d_{j-1} + d_j \qquad (7.3)$$

$$= a_j^{(1)} x_{j-2} + d_j^{(1)} \qquad (7.4)$$

Note that equation (7.4) is a linear first-order recurrence relation. It represents an improvement over the original equation (7.1), because it involves alternate terms of the original sequence. This process may be repeated $\log n$ times. After each step the following set of recurrence relations appears:

$$x_j = a_j^{(l)} x_{j-2^l} + d_j^{(l)}, \quad \text{where } l = 0, 1, \ldots, \log n; \ j = 1, 2, \ldots, n \qquad (7.5)$$

and where

$$a_j^{(l)} = a_j^{(l-1)} a_{j-2^{l-1}}^{(l-1)} \qquad (7.6)$$

$$d_j^{(l)} = a_j^{(l-1)} d_{j-2^{l-1}}^{(l-1)} + d_j^{(l-1)} \qquad (7.7)$$

with initial conditions

$$a_j^{(0)} = a_j \quad \text{and} \quad d_j^{(0)} = d_j$$

Whenever a subscript i of a_i, d_i, or x_i is found to be outside of the range 1 through n, then the value being referenced is 0. When $l = \log n$, the subscript of every reference to x_{j-2^l} in equation (7.5) is less than 1. Hence the solution to the recurrence relation can be found in $\log n$ reductions, since

$$x_j = d_j^{\log n} \qquad (7.8)$$

The key to the parallel algorithm is the parallel evaluation of equations (7.6) and (7.7) for all a_j and d_j. Equation (7.7) may look like a first-order linear recurrence, but it is not. The values computed at iteration l depend solely on values computed at iteration $l - 1$. Therefore, all the values on the lth iteration may be calculated simultaneously.

The parallel cyclic reduction algorithm appears below.

CYCLIC REDUCTION (SIMD-CC):
```
begin
    for i ← 0 to log n − 1 do
        for all Pj, where 2^i + 1 ≤ j ≤ n − 1 do
            if i ≥ 1 then
                t(j) ⇐ a(j − 2^i)
                a(j) ← a(j) + t(j)
            endif
            t(j) ⇐ d(j − 2^i)
            d(j) ← a(j) × t(j) + d(j)
        endfor
    endfor
    for all Pj do
        x(j) ← d(j)
    endfor
end
```

7-2 SOLVING PARTIAL DIFFERENTIAL EQUATIONS ON PROCESSOR ARRAYS

Methods to solve partial differential equations (PDEs) can be placed into one of two categories: direct methods and iterative methods. Direct methods solve the PDE analytically; iterative methods begin with guessed values at certain specified locations and then progress toward more accurate estimates. Iterative algorithms to solve PDEs can consume a great deal of computer time; hence they are a locus for interest in parallel computing. When the utility of an iterative method is evaluated, the rate of convergence of the values toward the solution is of great importance. Hockney and Jesshope [1981] are the primary reference for this section.

Consider the two-dimensional steady state temperature distribution problem illustrated in Figure 7-2. A thin steel plate is surrounded on three sides by a condensing steam bath (temperature 100°C). The fourth side is touching an ice bath (temperature 0°C). An insulating blanket covers the top and the bottom of the plate. The problem is to find the steady state temperature distribution at 100 evenly spaced points forming a 10×10 mesh in the plate.

This problem is an example of a linear second-order PDE. When the steady state temperature distribution has been found, a set of difference equations relates the values of variables on neighboring mesh points:

$$\phi_{x,y} = \frac{\phi_{x-1,y} + \phi_{x,y-1} + \phi_{x+1,y} + \phi_{x,y+1}}{4}$$

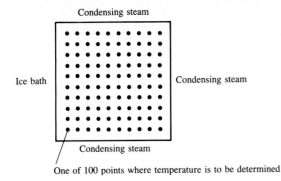

Condensing steam

Ice bath

Condensing steam

Condensing steam

One of 100 points where temperature is to be determined

Figure 7-2 Two-dimensional steady state temperature distribution problem.

Here the subscripts x and y refer to the coordinates of the mesh points. This problem is so simple that it can be solved analytically; however, more complicated PDEs must be solved iteratively. For that reason an iterative solution to the problem is explored.

Iterative procedures begin by assigning a guessed value for each variable $\phi_{x,y}$. The difference equation is then used to calculate successive values. The procedure is successful if the values of the variables converge to a solution.

Jacobi [1845] introduced a simple iterative procedure for updating the values of the variables ϕ. In his procedure the values at all the mesh points are simultaneously updated according to the formula

$$\phi'_{x,y} = \frac{\phi_{x-1,y} + \phi_{x,y-1} + \phi_{x+1,y} + \phi_{x,y+1}}{4}$$

The values of the ϕ values on the right-hand side of the equation are the old values of the variables; the ϕ' values represent the new values. The method proposed by Jacobi is obviously well suited for solution on a parallel computer, especially one based on the SIMD-MC^2 model, because all the variables that must be accessed by a processor are available in an adjacent processor.

If we assume this algorithm converges upon a solution, how many iterations must take place before the values of the variables stabilize? Unfortunately this is a difficult question to answer for the general case. Analytical results have been found for solving problems similar to our example, however. It has been shown that $en^2/2$ iterations are required in order to reduce the error by a factor of 10^{-e}. For example, to reduce the error by 10^{-3} on a 64×64 mesh would require about $3(64)^2/2 = 6144$ iterations, a slow rate of convergence.

The most popular iterative method for solving PDEs on sequential computers is successive overrelaxation (SOR) by points. The SOR differs from the Jacobi method in two important respects. First, a weighted average of $\phi'_{x,y}$ and $\phi^{\text{old}}_{x,y}$ is taken to determine the "new" value of the variable, $\phi_{x,y}$. Second, new values replace old values as soon as they are computed. In the typical sequential algorithm mesh points are processed one at a time, row by row. Replacing old values with new values as soon as they are computed greatly increases the rate of convergence. However, this method suffers from the disadvantage

E	O	E	O	E	O	E	O	E	O
O	E	O	E	O	E	O	E	O	E
E	O	E	O	E	O	E	O	E	O
O	E	O	E	O	E	O	E	O	E
E	O	E	O	E	O	E	O	E	O
O	E	O	E	O	E	O	E	O	E
E	O	E	O	E	O	E	O	E	O
O	E	O	E	O	E	O	E	O	E
E	O	E	O	E	O	E	O	E	O
O	E	O	E	O	E	O	E	O	E

Figure 7-3 Odd-even ordering.

that only one variable at a time is updated.

Fortunately, a parallel variant of this algorithm retains its high rate of convergence. The parallel algorithm is referred to as *odd-even ordering with Chebyshev acceleration*. Imagine the processing elements as forming a checkerboard. Every iteration has two phases. In the first phase all the "odd" processors come up with new values for their respective variables. In the second phase all the "even" processors generate new values for their variables (see Figure 7-3). The parallel algorithm also changes the weighting between $\phi'_{x,y}$ and $\phi^{old}_{x,y}$ at each half iteration.

The convergence rate for both SOR methods is identical. To reduce the error by a factor of 10^{-e}, the algorithms must perform approximately $ne/3$ iterations. For example, a 10^{-3} error reduction on a 64×64 mesh would require about $64(3)/3 = 64$ iterations, a great improvement over the more than 6000 iterations required by the Jacobi algorithm.

To contrast Jacobi's algorithm and odd-even ordering with Chebyshev acceleration, let us return to our example. Figures 7-4 and 7-5 illustrate the values of the variables at various iterations of Jacobi's algorithm and odd-even ordering with Chebyshev acceleration. Notice that the rate of convergence of the second algorithm is much higher.

7-3 SOLVING PARTIAL DIFFERENTIAL EQUATIONS ON MULTIPROCESSORS

Baudet [1978a, 1978b], Raskin [1978], and Deminet [1982] have implemented algorithms to solve PDEs on the Cm*. In this section we summarize the results of experiments reported by Deminet.

Deminet solves Laplace's equation with Dirichlet boundary conditions by the method of finite differences. The equation

$$\frac{\partial^2 z}{\partial x^2} + \frac{\partial^2 z}{\partial y^2} = 0$$

is solved for points on a 150×150 mesh by using iterative techniques similar to those described in the previous section. On every iteration the new value of each element is assigned the average of the values of its neighbors.

Each process, running on its own CPU, finds the values for a contiguous

Original estimates										After iteration 1									
91	81	81	54	98	97	55	86	83	69	60	81	83	93	80	77	92	81	81	83
57	51	98	95	69	57	84	85	69	50	49	83	70	80	84	82	71	77	66	66
54	96	55	98	87	77	86	69	47	25	58	64	87	81	77	79	75	65	47	59
77	95	56	87	63	87	69	44	27	39	58	82	84	68	87	70	64	49	44	52
85	98	98	57	88	69	40	31	46	57	64	83	77	90	64	62	51	46	43	56
81	54	97	86	69	34	37	54	58	39	59	93	79	73	58	55	47	41	40	67
96	98	77	69	25	46	59	39	22	53	69	71	83	55	60	39	40	43	49	42
98	57	69	31	57	39	37	36	47	7	63	83	53	59	37	47	46	43	19	63
98	69	46	39	53	47	49	49	9	50	59	60	56	45	48	37	37	28	46	40
69	39	47	49	50	7	15	20	39	43	59	71	58	59	52	53	44	51	43	72

After iteration 2										After iteration 3									
57	81	86	86	89	88	82	87	83	87	58	77	88	89	88	87	89	84	86	90
50	66	83	82	80	78	81	71	68	77	42	73	77	82	83	81	75	75	72	80
43	77	75	78	83	76	70	62	59	66	44	64	79	81	76	76	70	64	61	75
51	72	78	86	70	73	61	55	48	65	41	71	79	74	79	67	63	54	56	70
50	79	84	71	74	60	55	46	47	66	47	70	77	79	66	63	54	50	53	68
57	73	81	70	63	52	47	44	50	60	43	75	73	71	61	55	48	47	46	71
48	82	64	69	47	51	44	43	36	70	48	62	76	57	59	46	46	39	53	61
53	61	70	47	54	40	42	34	50	50	39	68	57	62	44	48	39	45	38	73
45	67	54	56	43	46	39	44	32	70	44	58	66	52	55	45	48	40	58	63
58	69	72	64	65	58	60	54	67	71	54	74	72	73	66	68	63	68	64	84

After iteration 40										After iteration 120									
51	71	81	87	90	93	94	96	97	99	51	72	82	87	91	94	95	97	98	99
32	53	67	75	82	86	89	92	95	97	32	54	67	77	83	88	91	94	96	98
23	43	56	67	74	81	85	89	93	97	24	43	58	69	77	83	88	92	95	98
20	36	51	61	70	76	82	87	92	96	20	38	52	64	73	80	86	90	94	97
18	34	47	59	67	75	80	86	90	95	18	35	50	62	71	79	85	89	93	97
18	34	48	58	68	74	81	85	91	95	18	35	50	62	71	79	85	89	93	97
19	37	50	61	69	77	81	87	91	96	20	38	52	64	73	80	86	90	94	97
23	42	56	66	74	79	85	88	93	96	24	43	58	69	77	83	88	92	95	98
32	53	66	75	81	86	88	92	94	97	32	54	67	77	83	88	91	94	96	98
51	71	81	86	90	92	94	96	97	99	51	72	82	87	91	94	95	97	98	99

Figure 7-4 Iterations of Jacobi's algorithm.

Original estimates

91	81	81	54	98	97	55	86	83	69
57	51	98	95	69	57	84	85	69	50
54	96	55	98	87	77	86	69	47	25
77	95	56	87	63	87	69	44	27	39
85	98	98	57	88	69	40	31	46	57
81	54	97	86	69	34	37	54	58	39
96	98	77	69	25	46	59	39	22	53
98	57	69	31	57	39	37	36	47	7
98	69	46	39	53	47	49	49	9	50
69	39	47	49	50	7	15	20	39	43

After iteration 1

60	82	83	112	80	81	92	88	81	101
44	83	71	80	88	82	79	77	67	66
58	62	87	61	77	75	75	56	47	101
29	82	97	68	75	70	55	49	66	52
64	63	77	82	64	53	51	59	43	73
36	93	69	73	58	55	55	41	44	67
69	69	83	69	60	54	40	47	49	84
16	83	71	59	50	47	45	43	53	63
59	66	56	69	48	46	37	41	46	87
48	71	92	59	78	53	97	51	91	72

After iteration 2

54	72	97	88	106	100	84	86	95	101
23	52	76	86	85	80	70	70	106	111
17	35	63	87	73	65	60	81	90	103
13	50	52	86	70	61	65	66	95	107
10	31	79	71	69	66	58	66	72	101
3	36	55	67	64	55	59	58	83	80
3	12	60	60	56	55	58	63	62	84
12	36	58	69	68	50	65	49	84	94
14	47	88	90	69	90	71	100	83	103
49	80	93	103	102	99	91	104	105	110

After iteration 3

45	66	76	86	86	84	92	103	101	99
32	51	64	83	80	80	95	95	94	98
18	41	62	63	79	83	76	93	100	96
13	23	42	61	69	70	79	83	92	95
10	17	37	48	67	68	68	87	95	95
15	17	28	60	64	64	71	74	92	99
6	29	38	57	65	75	62	89	88	101
21	30	46	69	84	80	93	95	102	92
34	50	63	75	98	93	95	97	106	97
52	68	74	93	93	94	99	96	97	97

After iteration 10

51	71	82	87	91	93	95	97	98	99
32	54	67	77	83	88	91	94	96	98
24	43	58	69	77	83	88	91	95	97
20	38	52	64	73	80	85	90	94	97
18	35	50	62	71	78	84	89	93	97
18	35	50	61	71	78	84	89	93	97
20	38	52	64	73	80	85	90	93	97
24	43	58	69	77	83	88	91	95	97
32	54	67	77	83	88	91	94	96	98
51	71	82	87	91	93	95	97	98	99

After iteration 15

51	72	82	87	91	94	95	97	98	99
32	54	68	77	83	88	91	94	96	98
24	43	58	69	77	83	88	92	95	98
20	38	52	64	73	80	86	90	94	97
18	35	50	62	71	79	85	89	93	97
18	35	50	62	71	79	85	89	93	97
20	38	52	64	73	80	86	90	94	97
24	43	58	69	77	83	88	92	95	98
32	54	68	77	83	88	91	94	96	98
51	72	82	87	91	94	95	97	98	99

Figure 7-5 Iterations of odd-even ordering with Chebyshev acceleration.

subsection of the mesh. To make the computation of indices as simple as possible, each process is assigned a number of complete rows and iteratively updates the values of the variables in these rows until the values stabilize.

Figure 7-6 illustrates the speedup achieved by the algorithm when the entire mesh is stored in a single cluster. The lower solid line represents the speedup achieved by the algorithm under the SMAP operating system. The upper solid line is the speedup that would have been achieved if all processes had finished every iteration as quickly as the fastest process. This line is straight over the

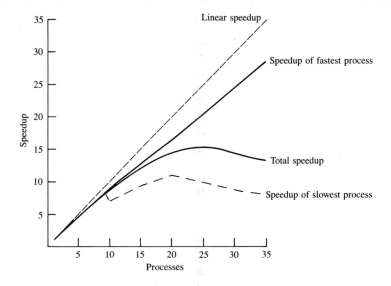

Figure 7-6 Speedup of PDE algorithm on Cm* (centralized data).

entire range of processes, because the process in the same computer module as the global data always has the fastest data access. The lower dashed line is the speedup that would have been achieved if all processes had finished every iteration as slowly as the slowest process. Notice that as long as all the processes are in the same cluster, the dashed and solid lines are virtually identical. However, as soon as processes are assigned to other clusters, they take longer to perform an iteration (because more time is spent accessing data), and the dashed line drops sharply. After this point the dashed line rises slowly, because processors are still being added, but soon it begins to decline, affecting the total speedup curve.

The first improvement Deminet makes to the algorithm is to do a better job of task selection. The original algorithm either assigned subsections to processors randomly or based the assignment on the computer module number. The improved algorithm assigns subsections that would seem to require more iterations—those in the middle of the mesh—to the faster processors—those in the same cluster as the data. As Figure 7-7 illustrates, this simple modification results in a 20 percent higher speedup for large numbers of processors.

The second modification is to distribute the data among the clusters containing active processors, with the goal of improving locality of reference. Figure 7-8 presents the results of this experiment, running under the SMAP operating system. The algorithm achieves much higher speedup than the original algorithm.

The dips in the speedup curve are a result of how the data are distributed. First, data are distributed to clusters, not computer modules. Each cluster has the same amount of data, even if it has fewer active computer modules. Second,

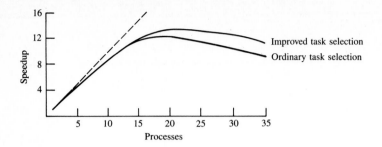

Figure 7-7 Speedup of PDE algorithm achieved with and without improved assignment of processes to tasks.

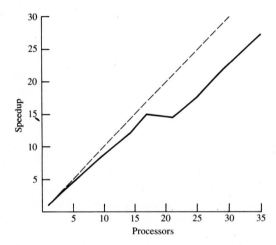

Figure 7-8 Speedup of PDE algorithm on Cm* (distributed data).

all the data of a particular cluster are stored in a single computer module. Thus a single computer module in each cluster has a much faster data access time. The algorithm achieves highest efficiency when the amount of data assigned to a cluster is roughly proportional to the number of active processors in that cluster, as is the case when $p = 17$.

7-4 GAUSSIAN ELIMINATION ON MULTIPROCESSORS

A system of linear equations $\mathbf{Ax} = \mathbf{b}$ can be solved in three steps: (1) Decompose $\mathbf{A}$ into the product of a lower triangular matrix $\mathbf{L}$ and an upper triangular matrix $\mathbf{U}$; (2) solve $\mathbf{Ly} = \mathbf{b}$; and (3) solve $\mathbf{Ux} = \mathbf{y}$ (Figure 7-9). Of these three steps, the triangularization procedure (step 1) has the highest computational complexity. In the following paragraphs we describe an efficient parallel triangularization procedure developed by Lord, Kowalik, and Kumar [1983].

Consider the sequential algorithm for performing an $\mathbf{LU}$ decomposition of

Three-step method to solve $\mathbf{Ax} = \mathbf{b}$ for $\mathbf{x}$.

$$
\begin{array}{c} \mathbf{A} \\ \begin{bmatrix} 3 & 2 & 1 & -2 \\ 9 & 4 & -5 & -4 \\ -6 & -3 & -4 & 12 \\ 3 & 4 & 3 & 9 \end{bmatrix} \end{array}
\begin{array}{c} \mathbf{x} \\ \begin{bmatrix} ? \\ ? \\ ? \\ ? \end{bmatrix} \end{array}
=
\begin{array}{c} \mathbf{b} \\ \begin{bmatrix} -7 \\ -11 \\ 36 \\ 22 \end{bmatrix} \end{array}
$$

STEP 1: Decompose $\mathbf{A}$ into LU.

$$
\begin{array}{c} \mathbf{L} \\ \begin{bmatrix} 1 & 0 & 0 & 0 \\ 3 & 1 & 0 & 0 \\ -2 & -.5 & 1 & 0 \\ 1 & -1 & 1 & 1 \end{bmatrix} \end{array}
\qquad
\begin{array}{c} \mathbf{U} \\ \begin{bmatrix} 3 & 2 & 1 & -2 \\ 0 & -2 & -8 & 2 \\ 0 & 0 & -6 & 9 \\ 0 & 0 & 0 & 4 \end{bmatrix} \end{array}
$$

STEP 2: Solve $\mathbf{Ly} = \mathbf{b}$ for $\mathbf{y}$ using forward substitution.

$$
\begin{array}{c} \mathbf{L} \\ \begin{bmatrix} 1 & 0 & 0 & 0 \\ 3 & 1 & 0 & 0 \\ -2 & -.5 & 1 & 0 \\ 1 & -1 & 1 & 1 \end{bmatrix} \end{array}
\begin{array}{c} \mathbf{y} \\ \begin{bmatrix} ? \\ ? \\ ? \\ ? \end{bmatrix} \end{array}
=
\begin{array}{c} \mathbf{b} \\ \begin{bmatrix} -7 \\ -11 \\ 36 \\ 22 \end{bmatrix} \end{array}
\qquad
\mathbf{y} =
\begin{bmatrix} -7 \\ 10 \\ 27 \\ 12 \end{bmatrix}
$$

STEP 3: Solve $\mathbf{Ux} = \mathbf{y}$ for $\mathbf{x}$ using back substitution.

$$
\begin{array}{c} \mathbf{U} \\ \begin{bmatrix} 3 & 2 & 1 & -2 \\ 0 & -2 & -8 & 2 \\ 0 & 0 & -6 & 9 \\ 0 & 0 & 0 & 4 \end{bmatrix} \end{array}
\begin{array}{c} \mathbf{x} \\ \begin{bmatrix} ? \\ ? \\ ? \\ ? \end{bmatrix} \end{array}
=
\begin{array}{c} \mathbf{y} \\ \begin{bmatrix} -7 \\ 10 \\ 27 \\ 12 \end{bmatrix} \end{array}
\qquad
\mathbf{x} =
\begin{bmatrix} 1 \\ -2 \\ 0 \\ 3 \end{bmatrix}
$$

Figure 7-9 Three-stage solution of a system of linear equations.

an $n \times n$ dense matrix $\mathbf{A}$ by using pivoting. To make the algorithm more legible, elements of $\mathbf{A}$ are denoted by parentheses, rather than subscripts.

```
LU DECOMPOSITION (SISD):
begin
    for k ← 1 to n − 1 do
        find l such that |a(l, k)| = max (|a(k, k)|, ... , |a(n, k)|)
        piv(k) ← l   {l is the pivot row}
        swap (a(piv(k), k), a(k, k))
        c ← 1/a(k, k)
        for i ← k + 1 to n do
            a(i, k) ← a(i, k) × c   {Elements of L}
        endfor
```

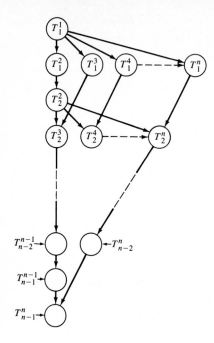

Figure 7-10 Maximally parallel graph upholding the precedence relations for **LU** decomposition. (Lord, Kowalik, and Kumar [1983]. Copyright © 1986 Association for Computing Machinery. Reprinted by permission.)

```
for j ← k + 1 to n do
  swap (a(piv(k),j), a(k,j))
  for i ← k + 1 to n do
    a(i,j) ← a(i,j) − a(i,k) × a(k,j)
  endfor
 endfor
 endfor
end
```

Define a *task* T_k^j to be the portion of the algorithm that computes a particular column of j, given some value of k. The set of tasks is denoted $J = \{T_k^j | 1 \le k \le j \le n, \ k \le n - 1\}$. Note that task T_k^j must be performed before task T_m^l if either (1) $j < l$ and $k = m$ or (2) $k < m$. The range and domain of these tasks are, respectively,

$$R(T_k^j) = \{a(i,j) | k \le i \le n\}$$
$$D(T_k^j) = \{a(i,j) | k \le i \le n\} \ \cup \ \{a(i,k) | k \le i \le n\}$$

Observe, for example, that the set of tasks

$$\{T_k^{k+1}, T_k^{k+2}, \ldots, T_k^n\}$$

are independent of each other and can be executed in parallel. Figure 7-10 is a maximally parallel graph that upholds all the precedence relations.

What is the minimum execution time of the **LU** decomposition, given the constraint graph of Figure 7-10? Assume that a time step consists of either a

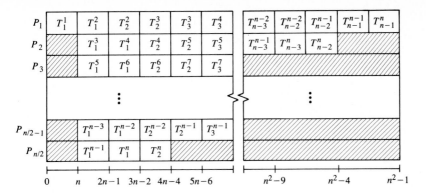

Figure 7-11 Gantt chart showing an optimal task schedule for $n = 10$ and $p = n/2 - 1 = 4$. (Lord, Kowalik, and Kumar [1983]. Copyright © 1986 Association for Computing Machinery. Reprinted by permission.)

multiply and a subtract or a multiply and a compare. Ignoring loop control overheads, we see the time needed to perform each task is

$$W(T_k^j) = \begin{cases} n + 1 - k & \text{if } k = j \\ n - k & \text{if } k < j \end{cases}$$

The longest path in the graph traverses the nodes

$$T_1^1, \; T_1^2, \; T_2^2, \; T_2^3, \; T_3^3, \; \ldots, \; T_{n-1}^{n-1}, \; T_{n-1}^n$$

in other words, the path down the left-hand side of the graph in Figure 7-10. The length of this path is

$$n + 1 + 2 \sum_{j=2}^{n-1} j = n^2 - 1$$

This path is so long that it is easy to come up with a schedule that assigns remaining tasks to $\lceil n/2 \rceil - 1$ processors so that no tasks finish after task T_{n-1}^n. Processor 1 is assigned the tasks along the critical path. Processor 2 is given tasks

$$T_1^3, \; T_1^4, \; T_2^4, \; T_2^5, \; T_3^5, \; \ldots, \; T_{n-2}^n$$

processor 3 is given tasks

$$T_1^5, \; T_1^6, \; T_2^6, \; T_2^7, \; T_3^7, \; \ldots, \; T_{n-4}^n$$

and processor j is given tasks

$$T_1^{2j-1}, \; T_1^{2j}, \; T_2^{2j}, \; T_2^{2j+1}, \; \ldots, \; T_{n-2(j-1)}^n$$

Note that these tasks are not paths through the graph. Figure 7-11 is a Gantt chart illustrating the task scheduling for even values of n.

The sequential **LU** decomposition algorithm takes $n^3/3 + O(n^2)$ time units to execute. Given p processors, the time needed to execute the parallel algorithm

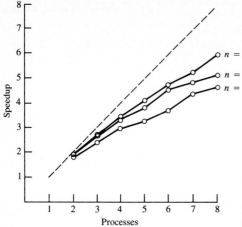

Figure 7-12 Speedup of parallel **LU** decomposition algorithm on Denelcor HEP.

is $(n^2 - 1)(n/2)/p$. Hence the efficiency of this algorithm using p processors is

$$E_p = \frac{n^3/3 + O(n^2)}{(n^2 - 1)n/2}$$

As $n \to \infty$, the efficiency of this algorithm approaches $\frac{2}{3}$. Even given relatively small values of n ($n \geq 50$), the algorithm is within 2 percent of its highest efficiency. Hence the algorithm seems amenable to parallelization: It can accommodate up to $n/2$ processors, and the processors are efficiently utilized.

Lord, Kowalik, and Kumar [1983] have implemented a gaussian elimination algorithm using this **LU** decomposition method on a single-PEM Denelcor HEP. The speedup achieved by the decomposition method is illustrated in Figure 7-12 for various matrix sizes.

7-5 SUMMARY

Recurrence relations often appear in the field of numerical analysis. The cyclic reduction algorithm allows linear first-order recurrences to be solved efficiently on the SIMD-CC model.

Partial differential equations appear frequently in engineering and the physical sciences. The SIMD-MC^2 model is capable of efficiently executing iterative algorithms for solving second-order differential equations in two or more dimensions. Deminet has developed a parallel algorithm to solve one kind of PDE on Cm*. His experiments illustrate once again the importance of locality of data references in the loosely coupled multiprocessor model.

Solving a system of linear equations $\mathbf{Ax} = \mathbf{b}$ for $\mathbf{x}$ is frequently performed in three stages. First, matrix $\mathbf{A}$ is decomposed into the product of a lower triangular matrix $\mathbf{L}$ and an upper triangular matrix $\mathbf{U}$. Second, the triangular system $\mathbf{Ly} = \mathbf{b}$ is solved for $\mathbf{y}$. Third, the triangular system $\mathbf{Ux} = \mathbf{b}$ is solved for **x**. Matrix decomposition is, in a sense, the inverse of matrix product, and solving a triangular system is similar to finding the inverse of matrix-vector product. Lord, Kowalik, and Kumar have implemented a parallel algorithm capable of efficiently performing the decomposition of a dense matrix.

BIBLIOGRAPHIC NOTES

Heller [1978] has surveyed parallel numerical algorithms to solve linear algebra problems. Hockney and Jesshope [1981] devote a chapter of their book to parallel algorithms. Topics they consider include the solution of recurrences, matrix multiplication, transforms, and partial differential equations.

Kogge and Stone [1973] introduced the recursive doubling algorithm for solving recurrence relations. Kung et al. [1982] describe the solution of partial differential equations on the Wavefront Array Processor.

References to **LU** decomposition in the literature include Chern and Murata [1983a], Horowitz [1979], Hwang and Cheng [1982], Jess and Kees [1982], and Kung et al. [1982]. Gilmore [1974] presents an algorithm to solve systems of linear equations on an associative processor. Chen and Wu [1984] also discuss the solution of dense linear systems of equations. LeBlanc [1986] examines the performance of two algorithms—one based on a shared-memory approach, the other using message passing—to perform gaussian elimination on the Butterfly multiprocessor.

Kumar and Kowalik [1984] have designed an MIMD algorithm to factor a positive definite matrix. Luk [1980] describes a singular-value decomposition algorithm for the ILLIAC IV.

Anybody seriously interested in developing parallel algorithms for matrix computations should be aware of Stewart's [1973] book, *Introduction to Matrix Computations*, because it describes many important sequential algorithms and shows how to analyze their effectiveness.

EXERCISES

7-1 How many iterations would be required by the Jacobi algorithm to improve the error bound by six decimal places on a 100×100 mesh? How many iterations would be required by the odd-even Chebyshev acceleration algorithm to achieve the same improvement on the same mesh?

7-2 Write an odd-even Chebyshev acceleration algorithm for the SIMD-MC^2 model. Do not include the test for convergence.

7-3 Discuss how a test for convergence of the odd-even Chebyshev acceleration algorithm could be performed on the SIMD-MC^2 model.

7-4 Predict the speedup that would be achieved by a tightly coupled multiprocessor executing the first PDE algorithm described for Cm*.

7-5 Describe how an upper triangular linear system can be transformed into a lower triangular linear system.

7-6 Explain why Lord, Kowalik, and Kumar's parallel **LU** decomposition algorithm takes $(n^2 - 1)(n/2)/p$ time units to execute, given p processors. (Assume n is an integer multiple of p.)

EIGHT

GRAPH ALGORITHMS

One way in which the early use of parallel computers has resembled the early use of sequential computers has been the emphasis on numerical algorithms. However, as the field of parallel algorithms matures, the emphasis can be expected to shift to nonnumerical algorithms, because more and more problems being solved on computers are symbolic in nature. This chapter examines a number of parallel algorithms developed to solve problems in graph theory. These problems relate to searching graphs and finding connected components, minimum spanning trees, and shortest paths in graphs.

Section 8-1 presents the graph theory terminology to be used in the rest of the chapter and outlines the various problems to be discussed; readers familiar with graph theory may wish to skip this section. Graph algorithms for SIMD models are covered in Section 8-2. The section begins with three SIMD-SM-R algorithms, analyzed by Reghbati and Corneil, for searching an undirected graph. Next we present Hirschberg's algorithm to solve the connected component problem. Adaptions of this algorithm to the SIMD-MC^2 and SIMD-P models are described. The section concludes by showing how the fast matrix multiplication algorithm of Dekel, Nassimi, and Sahni can be used to solve the all-pairs shortest-path problem on the SIMD-CC and SIMD-PS models. Section 8-3 presents tightly coupled multiprocessor algorithms for the minimum spanning tree and single-source shortest-path problems. This section illustrates how modifying a parallel algorithm to use a different data structure can improve its speedup. The primary references for this chapter are Quinn and Deo [1984] and Quinn and Yoo [1984].

8-1 TERMINOLOGY

A **graph** $G = (V, E)$ consists of V, a finite set of **vertices**, and E, a finite set of **edges** between pairs of vertices. In an **undirected graph** the edges have no orientation; in a **directed graph** every edge is an ordered pair (u, v) and is said to go **from** u **to** v. A **weighted graph** has a real number assigned to each edge. This number is called the **weight** of that edge, although depending upon the context it may be more appropriate to think of the number as a length, a time, a probability, or some other attribute.

The number of vertices in a graph is referred to by the letter n, and m denotes the number of edges in a graph. Four graph representations are common. An unweighted graph can be represented by an $n \times n$ matrix, with one row and one column for each vertex. The element at row i and column j is equal to 1 if and only if there is an edge from vertex i to vertex j; the value is 0 otherwise. This matrix is called an **adjacency matrix**. Weighted graphs can be represented by a **weight matrix**, which is similar to an adjacency matrix, except that the value of the matrix element at row i and column j gives the weight of the edge from vertex i to vertex j. Nonexistent edges may be represented by either 0 or ∞ entries, depending upon the particular problem being solved.

Second, a graph may be represented simply by a list of edges and a cardinal number indicating the number of vertices. A third representation uses adjacency lists—a list, for every vertex, of the edges leaving that vertex. A fourth representation, often used to represent digitized picture input, consists of a two-dimensional boolean matrix. If we label the elements 1 and 0, then the set of vertices consists of the matrix elements having the value 1, and the set of edges is the set of all pairs of vertically or horizontally adjacent 1s.

A **path** from v_1 to v_i in a graph $G = (V, E)$ is a sequence of edges (v_1, v_2), (v_2, v_3), (v_3, v_4), $\ldots$, (v_{i-2}, v_{i-1}), (v_{i-1}, v_i) such that every vertex is in V, every edge is in E, and no two vertices are identical. The definition for **cycle** is identical to that for path, except that the first and the last vertices are identical. A graph without cycles is said to be **acyclic**.

There are two common shortest-path problems. Given a weighted, directed graph $G = (V, E)$, the all-pairs shortest-path problem is to find, for every pair of vertices $i, j \in V$, the shortest path from i to j along edges in E. Given a weighted, directed graph $G = (V, E)$ and a vertex $s \in V$, the single-source shortest-path problem is to find, for every vertex $i \in V$, the shortest path from s to i along edges in E. Figure 8-1 illustrates these two kinds of shortest-path problems.

A **subgraph** of a graph G is a graph whose vertices and edges are in G. An undirected graph is **connected** if for every pair of vertices i and j in G, there is a path from i to j. The **connected component** problem is to find, for some undirected graph G, the minimal set of subgraphs such that every subgraph is connected (Figure 8-2). This problem is also known as the **component labeling** problem, since by the end of the algorithm every component's vertices share a label unique to that component. The **connected 1s** problem is the con-

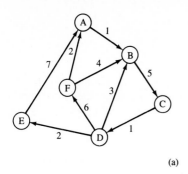

Length of shortest path

	A	B	C	D	E	F
A	0	1	6	7	9	13
B	14	0	5	6	8	12
C	9	4	0	1	3	7
D	8	3	8	0	2	6
E	7	8	13	14	0	20
F	2	3	8	9	11	0

(a)

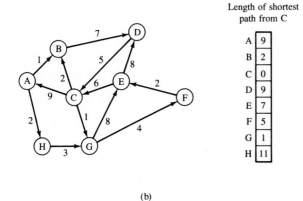

Length of shortest
path from C

A	9
B	2
C	0
D	9
E	7
F	5
G	1
H	11

(b)

Figure 8-1 Shortest-path problems. (a) All pairs. (b) Single source.

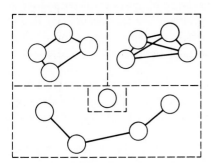

Figure 8-2 Connected components.

nected component problem applied to digitized picture input (the fourth graph representation described earlier).

A **tree** is a connected, undirected acyclic graph. A **spanning tree** for a graph G is a subgraph of G that is a tree containing every vertex of G. The

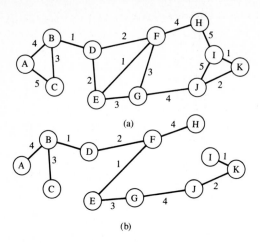

Figure 8-3 Minimum spanning tree problem. (a) Weighted graph. (b) Minimum spanning tree.

weight of a spanning tree of a weighted graph is the sum of the weights of its edges. Given a weighted, undirected graph G, a **minimum spanning tree** of G is a spanning tree with the smallest possible weight among all spanning trees of G (see Figure 8-3).

8-2 ALGORITHMS FOR PROCESSOR ARRAYS

Searching a Graph

Given an SIMD-SM-R model of computation with p processing elements, Reghbati (Arjomandi) and Corneil [1978] have determined the number of operations required for three parallel graph searching algorithms. Depth-first search seems to be an inherently sequential process, because searching always occurs along a single edge from a single vertex, restricting opportunities for parallelism. Reghbati and Corneil thus consider a parallel variant of depth-first search, called *p-depth search* (defined later), as well as parallel breadth-depth and parallel breadth-first search.

For these algorithms an adjacency matrix is not a suitable representation of the graph to be searched, since the process of searching through the elements of the matrix to find edges consumes too much processor time. Hence Reghbati and Corneil use adjacency lists to represent the graph.

How quickly can a graph be searched? Initially a master list of vertices still to be searched contains a single vertex. In any search procedure one vertex is chosen from the list of vertices to be searched. Each processor examines one or more edges emanating from that vertex. If the edge leads to a previously undiscovered vertex, it is added to that processor's partial list containing vertices to be added to the master list. At certain intervals the partial lists formed by the processors are linked and combined with the master list. Assume that the only operations that consume time are the vertex selection process and the list linking and combining process. Assume that it takes one of these *active operations* to select a vertex. For the sequential algorithm, only one active operation is required for the lone processor to add a new vertex to the master list. Let d_i denote the degree of vertex i. It is clear, then, that an upper bound for a sequential algorithm to search a graph is

$$T_1 = \sum_{i=1}^{n}(d_i + 1) = 2m + n$$

since vertex i can be added to a partial list only once, and d_i is the maximum number of times that vertex i can be chosen as the vertex from which searching is to be done.

P-Depth Search. In p-depth search, p edges incident upon a selected vertex are simultaneously searched. (In other words, processors are assigned to edges, one processor per edge.) One of the most recently searched vertices is then chosen as the point to continue the search. This procedure ends when the master list of vertices having unexplored edges is empty. Figure 8-4b illustrates p-depth search for $p = 2$. Note that if $p = 1$, the result is a depth-first search.

Theorem 8-1 (See Reghbati and Corneil [1978].) Given $p \geq 2$ processors, an upper bound on the number of active operations required by p-depth search on the SIMD-SM-R model is

$$\sum_{i=1}^{n} \left\lceil \frac{d_i + 1}{p} \right\rceil (\lceil \log p \rceil + 1)$$

Proof. The new vertices found at each state of the algorithm can be added to the master list of vertices in $\lceil \log p \rceil + 1$ active operations, by linking lists in a treelike manner. The number of times that searching begins at vertex i is $\lceil (d_i + 1)/p \rceil$. Therefore, the number of active operations required for parallel p-depth search is

$$T_p^1 = \sum_{i=1}^{n} \left\lceil \frac{d_i + 1}{p} \right\rceil (\lceil \log p \rceil + 1)$$

$$\leq \sum_{i=1}^{n} \left(\frac{d_i + 1}{p} + 1 \right) (\lceil \log p \rceil + 1)$$

$$\leq \frac{T_i(\lceil \log p \rceil + 1)}{p} + n(\lceil \log p \rceil + 1)$$

Note that in order for p-depth search to require fewer active operations than a sequential search, the term $(\lceil \log p \rceil + 1)/p$ must be less than 1. Hence $p \geq 4$ is a necessary condition for p-depth search to require fewer active operations than a sequential search. ∎

Graph:

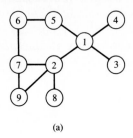

(a)

Two-processor p-depth search:

Parenthesized numbers indicate
order of traversal.

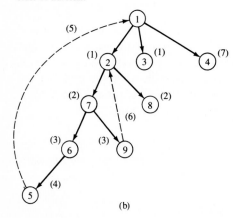

(b)

Figure 8-4 Parallel algorithms to search graphs. (a) Graph to be searched. (b) Two-processor p-depth search. (Quinn and Deo [1984]. Copyright © 1986 Association for Computing Machinery. Reprinted by permission.)

Breadth-Depth Search. A breadth-depth search proceeds by examining all the edges adjacent to a vertex before selecting one of the most recently reached vertices and continuing the search from that vertex. In parallel breadth-depth search, each processor keeps track of the new vertices it has discovered. Once all the edges from a vertex have been examined, these partial lists are linked and added to the master list. Since this parallel algorithm requires partial lists to be linked and combined with the master list less often than p-depth search, the algorithm requires fewer active operations. Figure 8-4c illustrates parallel breadth-depth search.

Theorem 8-2 (See Reghbati and Corneil [1978].) Given $p \geq 2$ processors, parallel breadth-depth search requires no more than

$$\sum_{i=1}^{n} \left(\left\lceil \frac{d_i}{p} \right\rceil + 1 + \lceil \log p \rceil + 1 \right)$$

active operations on the SIMD-SM-R model.

Proof. It requires $\lceil \log p \rceil + 1$ active operations to link the vertices found dur-

Two-processor breadth-depth search:

Parenthesized numbers indicate
order of traversal.

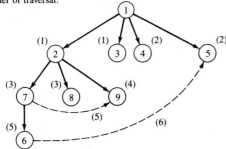

(c)

Two-processor breadth-first search:

Parenthesized numbers indicate
order of traversal.

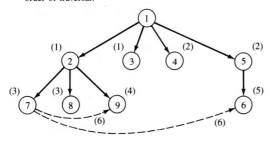

(d)

Figure 8-4 (*continued*) Parallel algorithms to search graphs. (c) Two-processor parallel breadth-depth search. (d) Two-processor parallel breadth-first search. (Quinn and Deo [1984]. Copyright © 1986 Association for Computing Machinery. Reprinted by permission.)

ing exploration from vertex i. There are $\lceil d_i/p \rceil + 1$ examination steps beginning from vertex i. Hence parallel breadth-depth search has an upper bound of

$$T_p^2 = \sum_{i=1}^{n} \left(\left\lceil \frac{d_i}{p} \right\rceil + 1 + \lceil \log p \rceil + 1 \right)$$

$$\leq \frac{T_1}{p} + n(\lceil \log p \rceil + 3) \quad \blacksquare$$

Breadth-First Search. Parallel breadth-first search requires even fewer link-and-combine steps, because processors examine all vertices at level i of the search tree before moving on to level $i + 1$ (Figure 8-4d). There is thus only one link-and-combine step for each level of the search tree.

Theorem 8-3 (See Reghbati and Corneil [1978].) Given $p \geq 2$ processors on the SIMD-SM-R model, the number of active operations required by parallel breadth-first search is

$$T_p^3 = \sum_{i=1}^{n} \left(\left\lceil \frac{d_i}{p} \right\rceil + 1 \right) + L \lceil \log p \rceil$$

where L is the distance of the vertex farthest from the start vertex.

Proof. This is left to the reader. ∎

Connected Components

There are three common approaches to find the connected components of an undirected graph. The first approach is to use some form of search, such as depth first or breadth first. The second approach is to solve the problem by finding the transitive closure of the adjacency matrix of the graph. Letting **A** denote the adjacency matrix of the original undirected graph G and **B** denote the transitive closure of **A**, we compute **B** by repeated ($\lceil \log n \rceil$) plus-min multiplications of **A**. The third approach collapses vertices into larger and larger sets of vertices until each set corresponds to a single connected component.

Hirschberg has used the third method to develop a connected components algorithm for processor arrays [Hirschberg 1976; Hirschberg, Chandra, and Sarwate 1979]. Although the algorithm is based on the unreasonable SIMD-SM-R model of parallel computation, it has been widely studied and merits discussion.

Theorem 8-4 (See Hirschberg [1976].) The connected components of an undirected graph can be found in $\Theta(\log^2 n)$ time on the SIMD-SM-R model with $\Theta(n^2)$ processors.

Summary of Algorithm. The primary data structure is the adjacency matrix. Instead of computing the transitive closure, however, adjacent vertices are combined into *supervertices*, which are themselves combined until each remaining supervertex represents a connected component of the graph. Like the transitive closure algorithm on the SIMD-SM-R model, this algorithm has a complexity of $O(\log^2 n)$, but requires only n^2 processors.

Each vertex is always a member of exactly one supervertex, and every supervertex is identified by its lowest-numbered member vertex, which is called the *root*. The parallel algorithm iterates through three stages. In the first, the lowest-numbered neighboring supervertex of each vertex is found. The second stage consists of connecting each supervertex root to the root of the lowest-numbered neighboring supervertex. In the third stage all newly connected supervertices are collapsed into larger supervertices. Since the number of supervertices is reduced by a factor of at least 2 in each iteration, $\lceil \log n \rceil$ iterations are sufficient to collapse each connected component into a single supervertex. The operation of this algorithm is illustrated in Figure 8-5. ∎

Theorem 8-5 (See Preparata and Probert [Hirschberg, Chandra, and Sarwate 1979].) The connected components of an undirected graph can be found in $\Theta(\log^2 n)$ time on the SIMD-SM-R model with $\Theta(n\lceil n/\log n\rceil)$ processors.

Summary of Algorithm. Hirschberg's original algorithm uses n^2 processors to assign values to the matrix containing the root numbers of neighboring supervertices. Preparata and Probert have observed that $\lceil n/\log n\rceil$ processors are sufficient to assign n values and find the minimum of n elements, both in $O(\log n)$ time, the reason being that each processor can assign values to $\log n$ elements, instead of to one element, without increasing the time complexity of the algorithm. Similarly, in the first phase of minimization, each processor can find the minimum of $\log n$ values, rather than two values, without increasing complexity of the algorithm. Hirschberg's algorithm thus can be implemented by using $n\lceil n/\log n\rceil$ processors, instead of n^2. ∎

Theorem 8-6 (See Chin, Lam, and Chen [1981, 1982].) The connected components of an undirected graph can be be found in $\Theta(\log^2 n)$ time on the SIMD-SM-R model with $\Theta(n\lceil n/\log^2 n\rceil)$ processors.

Summary of Algorithm. Every vertex is involved during each iteration of Hirschberg's algorithm. Chin, Lam, and Chen noted that by restricting participation in each iteration to a representative vertex of each supervertex (the root) and by removing isolated supervertices from further consideration, the algorithm requires fewer processors. ∎

Nassimi and Sahni [1980b] have adapted Hirschberg's algorithm to the SIMD-MC model. Let us examine their algorithm for the particular case of a two-dimensional mesh of processors. At various points in the parallel algorithm the processors execute a random access read or a random access write, as described in Chapter 4. Procedure **RANDOM_ACCESS_READ** has two arguments. The first argument denotes the address to receive the data; the second argument is an expression denoting the value to be read. For example, after statement **RANDOM_ACCESS_READ** $(a(i), b(c(i)))$ has executed, the value of local variable $a(i)$ of every unmasked processor $P(i)$ is identical to the value of variable $b(c(i))$, which is local to processor $P(c(i))$.

Procedure **RANDOM_ACCESS_WRITE** has two arguments. The first argument denotes the value to be written; the second argument is the address where the value is to be written. For example, when the statement

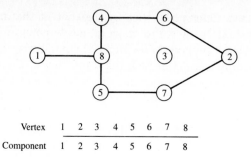

Vertex	1	2	3	4	5	6	7	8
Component	1	2	3	4	5	6	7	8

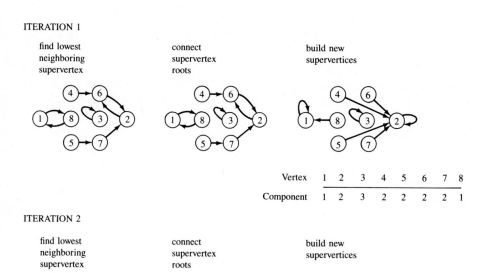

ITERATION 1

find lowest neighboring supervertex   connect supervertex roots   build new supervertices

Vertex	1	2	3	4	5	6	7	8
Component	1	2	3	2	2	2	2	1

ITERATION 2

find lowest neighboring supervertex   connect supervertex roots   build new supervertices

Vertex	1	2	3	4	5	6	7	8
Component	1	1	3	1	1	1	1	1

Figure 8-5 Hirschberg's connected component algorithm. (Quinn and Deo [1984]. Copyright © 1986 Association for Computing Machinery. Reprinted by permission.)

RANDOM_ACCESS_WRITE $(a(i), b(c(i)))$ executes, the value of local variable $a(i)$ of every unmasked processor $P(i)$ is written to variable $b(c(i))$ of processor $P(c(i))$. If more than one value is written to the same location, then the minimum of these values is the resultant value.

Recall that the complexity of performing a random access read or a random access write on the SIMD-MC^2 model with p processing elements is $O(p)$.

The algorithm uses a new function called BITS. BITS (i,j,k) returns the value of bits j through k of integer value i, where bit 0 is considered to be the least significant bit. For example,

$$\text{BITS }(17,3,1) = 0 \quad \text{BITS }(10,3,2) = 2$$
$$\text{BITS }(16,4,4) = 1 \quad \text{BITS }(15,2,3) = 0$$

Given a graph G with $n = 2^k$ vertices, the following SIMD-MC^2 algorithm finds the connected components of G. Assume that the maximum degree of any vertex in G is d. Let $adj(i,j)$, where $1 \leq i \leq n$ and $1 \leq j \leq d$, be the adjacency list of vertex i. If vertex i has $d_i < d$ edges, then $adj(i,j) = \infty$ for $d + i + 1 \leq j \leq d$. This adjacency list is stored in the local memory of processing element $P(i)$. Assume that $p = 2^k$ and that the processing elements are numbered in shuffled row-major order.

In addition to the edges of vertex i, each processing element $P(i)$ has a local variable $r(i)$, corresponding to the pointer to the supervertex root in Hirschberg's algorithm, and another local variable $temp(i)$, used as temporary storage. Procedure REDUCE corresponds to the third stage of Hirschberg's algorithm, collapsing vertices so that every vertex in a supervertex points to the root.

```
    REDUCE (r, n):
1.  begin
2.     for b ← 1 to log n do
3.        for all P(i) do
4.           if BITS(r(i), log n − 1, b) = BITS(i, log n − 1, b) then
5.              RANDOM_ACCESS_READ (r(i), r(r(i)))
6.           endif
7.        endfor
8.     endfor
9.  end
```

```
    CONNECTED COMPONENTS (SIMD-MC²):
1.  begin
       {Initially every node is a tree unto itself}
2.     for all P(i) do
3.        r(i) ← i
4.     endfor
       {Merge trees}
5.     for b ← 0 to ⌈log n⌉ − 1 do
          {It is not known whether there is a neighboring tree}
6.        for all P(i) do
7.           candidate(j) ← ∞
8.        endfor
          {Look for the lowest-numbered neighboring root}
9.        for e ← 1 to d do
```

```
10.          for all P(i) do
                 {adj(i, e) is the neighboring vertex;}
                 {r(adj(i, e)) is its root}
11.              RANDOM_ACCESS_READ (temp(i), r(adj(i, e)))
12.              if temp(i) = r(i) then
13.                  temp(i) ← ∞
14.              endif
15.              candidate(j) ← min(candidate(j), temp(j))
16.          endfor
17.       endfor
18.       for all P(i) do
                 {Each supervertex root gets the minimum
                    of the root numbers of neighboring supervertices}
19.              RANDOM_ACCESS_WRITE (candidate(i), r(r(i)))
                 {Take care of supervertices with no neighbors}
20.              if r(i) = ∞ then
21.                  r(i) ← i
22.              endif
                 {Make sure no cycles are formed}
23.              if r(i) > i then
24.                  RANDOM_ACCESS_READ (r(i), r(r(i)))
25.              endif
26.          endfor
                 {Collapse supervertices}
27.          REDUCE (r, n)
28.       endfor
29.  end
```

Theorem 8-7 (See Nassimi and Sahni [1980b].) The connected components of an undirected graph G with $n = 2^k$ vertices and maximum vertex degree d can be found in time $O(dn \log n)$ on the SIMD-MC^2 model having n processors.

Complexity Analysis. This is left to the reader. ∎

Theorem 8-8 (See Miller and Stout [1986].) Suppose the adjacency matrix of an undirected graph G with n vertices is stored in the base of a SIMD-P (pyramid) computer of size n^2. Then the connected components of G can be found in time $O(\sqrt{n})$.

Miller and Stout, like the authors previously mentioned, have based their algorithm on the work of Hirschberg. Their innovation is their effective use of the topology of the pyramid. As the algorithm progresses, there are fewer and fewer supervertices (min trees) to be combined. When the forest of min trees is to be relabeled, data are moved up the pyramid, where the combining step can be performed on a mesh of appropriate size.

Determining the minimum spanning forest of a weighted graph can be seen as a simple variation of determining connected components. At each iteration the minimum edge, rather than the minimum labeled vertex, is found. Hence similar complexity results exist for the minimum spanning forest problem. For example, consider Theorem 8-9.

Theorem 8-9 (See Miller and Stout [1986].) Suppose the weight matrix of a weighted, undirected graph G with n vertices is stored in the base of an SIMD-P (pyramid) computer of size n^2. Then the minimum spanning forest of G can be found in time $O(\sqrt{n})$.

All-Pairs Shortest Path

The fast matrix multiplication algorithm of Dekel, Nassimi, and Sahni [1981], first studied in Chapter 6, can be used to solve a number of graph problems efficiently, including the all-pairs shortest-path problem.

Theorem 8-10 (See Dekel, Nassimi, and Sahni [1981].) Given an n-vertex weighted graph, the all-pairs shortest-path problem can be solved in $\Theta(\log^2 n)$ time on the SIMD-CC and SIMD-PS models, given $n^3 = 2^{3q}$ processors.

Proof. Let G be an n-vertex weighted graph. Our goal is to produce an $n \times n$ matrix $\mathbf{A}$ such that $a_{i,j}$ is the length of the shortest path from i to j in G. Let $a_{i,j}^k$ denote the length of the shortest path from i to j with at most $k-1$ intermediate vertices. Since there are no negative weight cycles in G, $a_{i,j} = a_{i,j}^{n-1}$. In this example, $a_{i,i}^1 = 0$, for all i, $1 \le i \le n$, and for all distinct i and j, $a_{i,j}^1$ is the weight of the edge from i to j; if no such edge exists, $a_{i,j}^1 = \infty$. It follows from the principle of combinatorial optimality that $a_{i,j}^k = \min_m \{ a_{i,m}^{\lceil k/2 \rceil} + a_{m,j}^{\lceil k/2 \rceil} \}$. Hence $\mathbf{A}^{n-1}$ may be computed from $\mathbf{A}^1$ by repeated plus-min multiplications. By substituting plus for multiply and min for plus, $\lceil \log n \rceil$ matrix multiplications are sufficient to generate the matrices $\mathbf{A}^2$, $\mathbf{A}^4$, $\ldots$, $\mathbf{A}^{n-1}$. Recall that a single matrix multiplication has complexity $O(\log n)$ on the SIMD-CC and SIMD-PS models, given $n^3 = 2^{3q}$ processors. Thus the all-pairs shortest-path problem can be solved in $\Theta(\log^2 n)$ time on the SIMD-CC and SIMD-PS models, given $n^3 = 2^{3q}$ processors. ∎

8-3 ALGORITHMS FOR MULTIPROCESSORS

Minimum Spanning Tree

Efforts to find the minimum spanning tree of a weighted, connected, undirected graph have focused on three classical sequential algorithms: Sollin's [1977] algorithm, the Prim-Dijkstra algorithm [Prim 1957; Dijkstra 1959], and Kruskal's [1956] algorithm. In this section parallel algorithms based on Sollin's algorithm and Kruskal's algorithm are described.

Sollin's Algorithm. The most obvious candidate for investigation is a sequential algorithm attributed to Sollin. In this algorithm one starts with a forest of n isolated vertices, with every vertex regarded as a tree. In an iteration, the algorithm simultaneously determines for each tree in the forest the smallest edge joining any given vertex in that tree to a vertex in some other tree. All such edges are added to the forest, with the exception that two trees are never joined by more than one edge. (Ties between edges, which would cause a cycle, are resolved arbitrarily.) This process continues until there is only one tree in the forest—the minimum spanning tree. Since the number of trees is reduced by a factor of at least 2 in each iteration, Sollin's algorithm requires at most $\lceil \log n \rceil$ iterations to find the minimum spanning tree. An iteration requires at most $O(n^2)$ comparisons to find the smallest edge incident on each vertex. Thus the sequential algorithm has complexity $O(n^2 \log n)$. Sollin's algorithm is illustrated in Figure 8-6. Pseudocode for the algorithm follows. Note that this algorithm uses sets to keep track of which vertices are in which trees. The **FIND** function, passed a vertex v, returns the name of the set (tree) v is in. The procedure **UNION**, passed two vertices v and w, performs the set union of the sets containing v and w; in other words, it connects the trees containing vertices v and w. Hopcroft and Ullman [1973] showed how these two operations can be performed extremely efficiently. Readers unfamiliar with these two operations can find descriptions in the above-mentioned reference or in a variety of texts on algorithms, including Aho, Hopcroft, and Ullman [1974].

```
    MINIMUM SPANNING TREE (SISD):
 1. begin
 2.     for i ← 1 to n do
 3.         Vertex i is initially in set i
 4.     endfor
 5.     T ← ∅                          {T is the minimum spanning tree}
 6.     while |T| < n − 1 do
 7.         for every tree i do
 8.             closest(i) ← ∞
 9.         endfor
10.         for every edge {v, w} do
11.             if FIND(v) ≠ FIND(w) then
12.                 if weight({v, w}) < closest(FIND(v)) then
13.                     closest(FIND(v)) ← weight({v, w})
14.                     edge(FIND(v)) ← {v, w}
15.                 endif
16.             endif
17.         endfor
18.         for every tree i do
19.             {v, w} ← edge(i)
20.             if FIND(v) ≠ FIND(w) then
21.                 T ← T ∪ {v, w}
22.                 UNION(v, w)
23.             endif
24.         endfor
25.     endwhile
26. end
```

How should this algorithm be parallelized for the tightly coupled multiprocessor model? According to the grain size lemma, parallelization should be done on the outermost loop possible. Unfortunately, we cannot make the while loop parallel, because there are precedence constraints between iterations. Each of the trees existing on iteration i must be joined with the nearest tree before iteration $i + 1$ can begin. Hence parallelization must be done inside the while loop. Lines 7 to 9 can be made parallel through the now familiar method of prescheduling: Each of the p processors is responsible for $1/p$ of the trees. The for loop in lines 10 to 17 can also be made parallel through prescheduling. This is most efficiently done by assigning each processor its fair share of the vertices, then allowing it to examine every outgoing edge from this set.

Parallelizing the loop in lines 18 to 24 is more complicated. The complicating situation is illustrated in Figure 8-7. Suppose one processor is attempting to connect tree A with its closest neighbor B, while another processor is attempting to connect tree B with its closest neighbor A. Variable $edge(A)$ contains edge $\{v_A, w_A\}$ with length k. Variable $edge(B)$ contains edge $\{v_B, w_B\}$, also with length k. If both processors perform the test in line 20 before either processor performs the UNION operation in line 22, then both edges will be added to T,

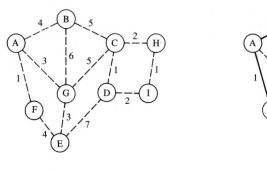
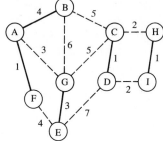
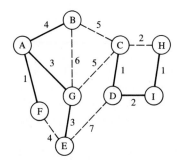
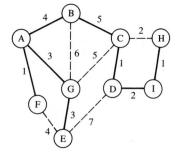

Figure 8-6 Sollin's minimum spanning tree algorithm.

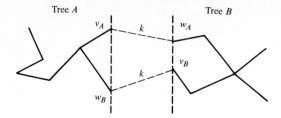

Figure 8-7 A complication that arises in the parallelization of Sollin's minimum spanning tree algorithm.

an error. Therefore, if the loop in lines 18 to 25 is to be made parallel, trees FIND(v) and FIND(w) must be locked before line 20 and unlocked after line 23, since we can allow only one processor at a time to operate in this critical section. Given that only one tree can be locked at a time, care must be taken to avoid deadlock. One way to prevent this is to lock the lower-numbered tree first.

Theorem 8-11 The parallel version of Sollin's algorithm described above has complexity $O(\lceil \log n \rceil (n^2/p + n/p + n + p))$.

Proof. A series of n **UNION** and **FIND** operations has worst-case time complexity $O(n \log^* n)$: the time spent per individual operation amortizes to $O(\log^* n)$, virtually a constant. Hence we make the reasonable simplifying assumption that **UNION** and **FIND** are constant-time operations. The parallelization of the **for** loop in lines 7 to 9 has complexity $O(n/p+p)$. The parallelization of the **for** loop in lines 10 to 17 has complexity $O(n^2/p + p)$. The parallelization of the **for** loop in lines 18 to 24 has complexity $O((n/p)p + p)$. The factor of p in the last loop occurs because in the worst case a processor wanting to lock a particular tree A may have to wait for every other processor to lock and unlock A. Remember that the outer **while** loop executes at most $\lceil \log n \rceil$ times. The overall complexity of the parallel version of Sollin's algorithm, then, is $O(\lceil \log n \rceil (n^2/p + n/p + n + p))$. ∎

The complexity of this parallel algorithm is minimized when $p = O(\sqrt{n})$. If $p \ll n$, we can expect the parallel algorithm to achieve good speedup.

Kruskal's Algorithm.   In Kruskal's algorithm the graph initially consists of a forest of isolated vertices. The edges are scanned in nondecreasing order of their weights, and every edge that connects two disjoint trees is added to the minimum spanning tree. (In other words, all edges that do not cause cycles with existing edges are selected.) The algorithm halts when the graph consists of a single tree, a minimum spanning tree. Figure 8-8 illustrates Kruskal's algorithm.

Lemma 8-1 (See Quinn and Yoo [1984], Yoo [1983].) A tightly coupled multiprocessor with $\lceil \log m \rceil$ processors can remove an element from an m-element heap in constant time.

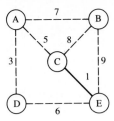
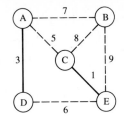

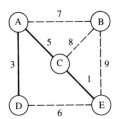
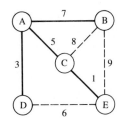

Figure 8-8 Kruskal's minimum spanning tree algorithm.

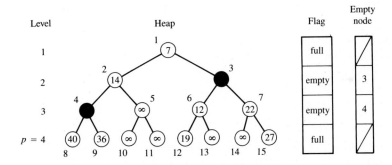

Figure 8-9 Yoo's pipelined algorithm to empty heap. (Quinn and Deo [1984]. Copyright ©
1986 Association for Computing Machinery. Reprinted by permission.)

Description of Algorithm. This pipelined algorithm is illustrated in Figure
8-9. An array is used to implement the heap in the usual way, with the root
of the heap being stored in the first element and the left and right children of
node i being stored in elements $2i$ and $2i + 1$. The heap is a full binary tree
with p levels, p being the number of processors. Some nodes in the bottom level
of the tree are assigned the value ∞, if necessary, to fill the tree. During the
course of an algorithm's execution, a node is *full* if it contains a value (including
∞). A node is *empty* if its value has been transmitted to its parent and no
replacement value has been received from any of its children. The array *flag*

indicates which levels contain an empty node; $flag(i) = empty$ if level i has an empty node, otherwise $flag(i) = full$. If $flag(i) = empty$, then $empty\_node(i)$ indicates which node is the empty one. For all i, where $2 \leq i \leq p$, processor i is assigned the task of keeping all the nodes at level $i-1$ full. If $flag(i-1) = empty$ and $flag(i) = full$, then processor i fills the empty node at level $i-1$ with the appropriate child at level i. Then $flag(i-1)$ becomes $full$, and $flag(i)$ becomes $empty$. When a leaf node is emptied, it is filled with an ∞. Eventually the ∞'s fill the tree. Processor 1 empties node 1 whenever it is full and terminates the procedure when node 1 has the value ∞ (i.e., when the heap is empty). ∎

Yoo has also described a parallel method for initializing a heap. The common sequential algorithm can be described as follows:

```
HEAP CONSTRUCTION (SISD):
1.  begin
2.     d ← depth of the heap
3.     for l ← d − 1 downto 0 do
4.        for each nonleaf node v at level l do
5.           k ← heap(v)
6.           repeat
7.              w ← child of v with the larger key
8.              if k < key(w) then
9.                 key(v) ← key(w)
10.                v ← w
11.             else exit repeat loop
12.             endif
13.          until v is a leaf node
14.          key(v) ← k
15.       endfor
16.    endfor
17. end
```

The **for** loop in lines 4 to 16 can be made parallel in a straightforward manner by using prescheduling; each processor makes into heaps its share of the nonleaf nodes at level l. A better method is to allow heap construction to occur at a node once both of its children have been made into heaps [Yoo 1983]. This method does away with the requirement that processors synchronize as they reach the end of every level.

Theorem 8-12 (See Yoo [1983].) The minimum spanning tree of a weighted, undirected, connected graph with m edges can be found in time $O(m)$ on a tightly coupled multiprocessor with $\lceil \log m \rceil$ processors.

Summary of Algorithm. The first step is to make a heap out of the edges of the weighted graph, based on their lengths. In the second step a single process repeatedly removes an edge from the heap and determines whether it is an element of the minimum spanning tree, while the remaining processes repeatedly restore the heap. Recall that the edge is an element of the minimum spanning tree if it connects two unconnected subtrees. Determining whether an edge connects two unconnected subtrees and connecting the subtrees, if necessary, can be accomplished in virtually constant time by using the efficient **FIND** and **UNION** algorithms described earlier. The pipelined processes emptying the heap can provide a new edge in constant time. Since the heap contains m edges, the worst-case time complexity is $O(m)$. ∎

Single-Source Shortest Path

The **single-source shortest-path problem** is to find the shortest path from a specified vertex s, the **source**, to all other vertices in a weighted, directed graph. Let $weight(u, v)$ represent the length of the edge from u to v; if no such edge exists, then $weight(u, v) = \infty$. In Moore's single-source shortest-path algorithm $distance(v)$ is initially assigned the value ∞, for all $v \in V - \{s\}$. The distance from s to itself is, of course, zero. A queue contains vertices from which further searching must be done; initially it contains s. As long as the queue remains nonempty, the vertex u from the head of the queue is removed, and all edges $(u, v) \in E$ are examined. If $distance(u) + weight(u, v) < distance(v)$, then a new shorter path to v has been found (the one through u). In this case $distance(v)$ is revised, and v is added to the tail of the queue, if it is not already in the queue. The algorithm continues this process until the queue is empty. Moore's algorithm is shown in Figure 8-10.

```
    SHORTEST PATH (SISD):
    global  distance,   {Element i contains distance from s to i}
            n,          {Number of vertices in graph}
            s,          {Source vertex}
            weight      {Contains weight of every edge}
 1. begin
 2.    for i ← 1 to n do
 3.       INITIALIZE(i)
 4.    endfor
 5.    insert s into the queue
 6.    while the queue is not empty do
 7.       SEARCH
 8.    endwhile
 9. end


    SEARCH:
    local  new_distance,  {Distance to v if pass through u}
           u,             {Examined edge leaves this vertex}
           v,             {Examined edge enters this vertex}
10. begin
11.    dequeue vertex u
12.    for every edge {u, v} in the graph do
13.       new_distance ← distance(u) + weight({u, v})
```

```
14.        if new_distance < distance(v) then
15.            distance(v) ← new_distance
16.            if v is not in the queue then
17.                enqueue vertex v
18.            endif
19.        endif
20.    endfor
21. end
```

Procedure **INITIALIZE**, called in line 4, initializes the distance of every non-source vertex to ∞ and the distance of s to zero. The for loop in lines 12 to 20 corresponds to the search for shorter paths to vertices directly reachable from vertex u.

It would seem that this algorithm is amenable to parallelization. There are two obvious methods to consider. The first method makes the for loop in lines 12 to 18 parallel. Any given vertex is likely to have several outgoing edges. These could all be explored in parallel. The second method is to parallelize the while loop in lines 6 to 8. At any one time in the execution of the algorithm there are probably many vertices in the queue. It should be possible to explore edges from more than one vertex at a time. Which method is better? There are at least two reasons to favor the second method. First, the second method produces larger-grained tasks for the processes to perform. Hence by the grain size lemma it would seem more likely to produce good speedup. Second, the parallelizability of the first method is limited by the number of edges leaving each vertex. If the graph is relatively sparse, i.e., relatively few edges per vertex, then the number of processes that can be used is too constrained.

Consider the following parallel algorithm, which is based on the second method described. The queue is initialized with the source vertex, and then a number of asynchronous processes are created. Each of these processes goes through the steps of deleting a vertex from the queue, examining its outgoing edges, and inserting into the queue the vertices to which shorter paths have been found.

The for loop in lines 2 to 4 of the sequential algorithm is easily transformed to a parallel for loop by using the method of prescheduling. The parallel for loop occupies lines 2 to 6 of the parallel algorithm shown below. The while loop of lines 6 to 8 must be changed to reflect the existence of a number of asynchronous processes performing the **SEARCH** procedure in parallel. Clearly it is not appropriate for a process to terminate when it discovers that the queue is empty. (Why?) Hence a more complicated method must be used. In the following algorithm two variables are used together to determine when there is no more work to do. The first variable, *waiting*, is an array keeping track of which processes are waiting for work to do. The second variable, the boolean *halt*, is set to the value *true* only when all the processes are waiting and the queue is empty. Procedure **INITIALIZE** initializes every entry in array *waiting* to *false*. Lines 6 to 8 of the sequential algorithm become lines 8 to 11 of the parallel algorithm.

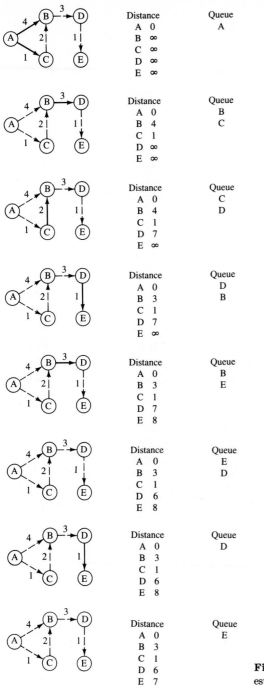

Distance		Queue
A	0	A
B	∞	
C	∞	
D	∞	
E	∞	

Distance		Queue
A	0	B
B	4	C
C	1	
D	∞	
E	∞	

Distance		Queue
A	0	C
B	4	D
C	1	
D	7	
E	∞	

Distance		Queue
A	0	D
B	3	B
C	1	
D	7	
E	∞	

Distance		Queue
A	0	B
B	3	E
C	1	
D	7	
E	8	

Distance		Queue
A	0	E
B	3	D
C	1	
D	6	
E	8	

Distance		Queue
A	0	D
B	3	
C	1	
D	6	
E	8	

Distance		Queue
A	0	E
B	3	
C	1	
D	6	
E	7	

Figure 8-10 Moore's single-source shortest-path algorithm.

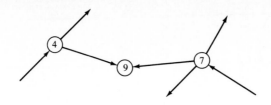

Current value of *distance* (9)	Process 1	Process 2
28	Dequeue vertex 4	Dequeue vertex 7
28	Consider edge (4,9)	Consider edge (7,9)
28	*new__distance* ← 22	*new__distance* ← 24
28	*new__distance* < *distance* (9)	*new__distance* < *distance* (9)
28	*distance* (9) ← 22	•
22	•	*distance* (9) ← 24
24	•	•

Figure 8-11 If locking is not used, two processes may attempt to update the value of *distance(v)* simultaneously, causing an error.

How must procedure **SEARCH** be modified? Because enqueuing and dequeuing are not atomic operations, whenever an element is enqueued or dequeued, the queue must be locked. Second, before a process compares the newly found distance to vertex v, *new_distance*, to the current shortest distance to v, *distance(v)*, variable *distance(v)* must be locked. Otherwise, two processes could both find themselves trying to update *distance(v)* simultaneously, and *distance(v)* could end up with the wrong value (see Figure 8-11). Finally, if a process finds that the queue is empty, then it sets its entry in array *waiting* to *true*. If process 1 is waiting, then it checks to see whether every process is waiting. If every process is waiting, the value of *halt* is set to *true*. Notice that the queue must be locked while process 1 checks to see whether every process is waiting. The parallel algorithm follows.

```
SHORTEST PATH (TIGHTLY COUPLED MULTIPROCESSOR):
global  distance,   {Element i contains distance from s to vertex i}
        halt,       {Set to true when it is time for processes to stop}
        n,          {Number of vertices in graph}
        p,          {Number of processes}
        s,          {Source vertex}
        weight      {Contains weight of every edge}
1. begin
2.   for all Pᵢ, where 1 ≤ i ≤ p do
3.      for j ← i to n step p do
4.         INITIALIZE(j)
5.      endfor
6.   endfor
```

```
 7.    enqueue s
 8.    halt ← false
 9.    for all i, 1 ≤ i ≤ p do
10.       repeat SEARCH (i) until halt
11.    endfor
12. end


        SEARCH (i):
        parameter   i                {Process number}
        local       new_distance,    {Distance to v if go through u}
                    u,               {Edge is directed from this vertex}
                    v                {Edge is directed to this vertex}
13. begin
14.     lock the queue
15.     if the queue is empty then
16.         waiting(i) ← true
17.         if i = 1 then
18.             halt ← waiting(2) and waiting(3) and ... and waiting(p)
19.         endif
20.         unlock the queue
21.     else
22.         dequeue u
23.         waiting(i) ← false
24.         unlock the queue
25.         for every edge {u, v} in the graph do
26.             new_distance ← distance(u) + weight({u, v})
27.             lock (distance(v))
28.             if new_distance < distance(v) then
29.                 distance(v) ← new_distance
30.                 unlock (distance(v))
31.                 if v is not in the queue then
32.                     lock the queue; enqueue v; unlock the queue
33.                 endif
34.             else unlock (distance(v))
35.             endif
36.         endfor
37.     endif
38. end
```

How much speedup is achievable by this algorithm? Initially, creating more processes decreases the total execution time of the algorithm, because the outgoing edges of several vertices can be examined in parallel. However, since each process demands exclusive control of the queue to insert or delete vertices, maximum speedup is eventually constrained.

There would be no contention between the processes if each process maintained a private list of vertices to be searched, inserting and deleting elements on its own. If the variance in the size of these lists were large, however, then letting each process handle its own list could cause a severe imbalance in the workloads. A middle course is to let each process insert elements into its own

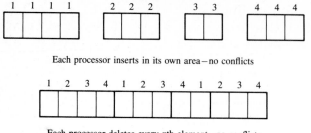

Each processor inserts in its own area—no conflicts

Each processor deletes every pth element—no conflicts

Figure 8-12 Logical form of linked array.

private space, then join these lists, so that deletions can occur by letting each of the p processes examine every pth element of the combined list, balancing the work done by each process (Figure 8-12).

The **linked array** is a data structure designed to allow the joining of various sized lists so that the inserting and searching of the list elements can be done in parallel without contention. Assume that in a single iteration no processor ever inserts more than w elements. The linked array, in that case, contains $p(w + p)$ elements, $w + p$ elements per process. Each contiguous group of $w + p$ locations is the space where a process may store the names of elements to be searched in the next iteration. If processor i, $1 \leq i \leq p$, generates the names of e_i elements to be considered in the next iteration, then locations $(i - 1)(w + p) + 1$ through $(i - 1)(w + p) + e_i$ contain these names. Locations $(i - 1)(w + p) + e_i + 1$ through $(i - 1)(w + p) + e_i + p$ contain the values $-i(w + p + 1)$ down to $-[i(w + p) + p]$, respectively.

In the next iteration, when the elements whose names are in the linked array are to be examined, processor i, $1 \leq i \leq p$, examines every pth location, beginning with location i. If the value encountered is greater than zero, it is the name of a vertex to be searched. If the value is less than zero, it is a pointer, and the processor immediately jumps to the index indicated (the absolute value). When the pointer has value less than $-p(w + p)$, the search terminates.

Theorem 8-13 (See Quinn [1983].) Given a set of n elements stored in a linked array and $p \geq 1$ processors, the difference between the greatest number of elements searched by any processor and the least number of elements is less than or equal to 1.

Proof. This is left to the reader. ∎

The following procedure illustrates how a process removes its share of elements from an array containing vertex numbers and pointers.

EXAMINE (a,i,p,w):

parameter	$a,$	{Array containing vertex numbers and pointers}
	$i,$	{Process number}
	$p,$	{Number of processes}
	w	{Size of subarray allocated to each process}
local	j	{Index into a}

```
 1. begin
 2.    j ← i
 3.    while j ≤ p(w + p) do
 4.       if a(j) < 0 then
 5.          j ← −a(j)                    {Follow pointer}
 6.       else
 7.          manipulate vertex whose value is a(j)
 8.          j ← j + p
 9.       endif
10.    endwhile
11. end
```

Linked arrays can be costly in terms of the space overhead. Unless it can be guaranteed that not all the insertions will be done by a single processor, then the space allocated to the linked array must be approximately p times the space allocated to a simple array. A partitioned parallel version of Moore's algorithm can be devised to use the linked array. In a single iteration each process has a number of vertices to examine. Every process compiles its own list of vertices to which shorter paths have been found and builds links to the next process's list. In the following iteration these lists are examined in parallel. Two arrays are used. In any iteration one array is being read while the other array is being written. In the next iteration the roles of the two arrays are reversed.

Figure 8-13 contrasts the speedup of the linked array version of Moore's algorithm with the speedup achieved by Deo, Pang, and Lord's [1980] sequential deque version of Pape-d'Esopo's algorithm (a variant of Moore's algorithm that uses a double-ended queue, or **deque**) on a single PEM HEP running under the HEP/OS operating system. Processes executing this new algorithm must spend time traversing links and synchronizing with each other, which adds to the overhead of the algorithm. Thus the slope of the speedup curve is less than that of the relaxed parallel algorithm that uses the sequential deque. However, the problem of software lockout is eliminated in this algorithm, and the maximum speedup achieved is higher.

8-4 SUMMARY

A number of graph algorithms have been presented in this chapter. Three parallel graph searching algorithms were defined, and their complexity was analyzed for the SIMD-SM-R model. Many algorithms have appeared to solve the connected components problem. Most of these algorithms are related to Hirschberg's

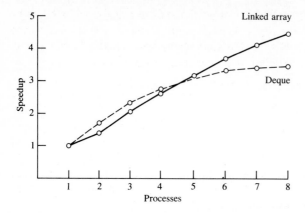

Figure 8-13 Speedup of parallel Moore-algorithms on single PEM Denelcor HEP.

algorithm, which uses the vertex-collapsing approach.

The efficient matrix multiplication algorithm developed by Dekel, Nassimi, and Sahni for the SIMD-PS and SIMD-CC models can be used to solve a variety of graph theoretic problems, including the all-pairs shortest-path problem.

We examined parallelizations of two minimum spanning tree algorithms for tightly coupled multiprocessors. Sollin's algorithm has a straightforward parallelization. Kruskal's algorithm also has an efficient parallelization for a small number of processors, because it is possible to pipeline the initialization and the emptying of a heap.

Deo, Pang, and Lord's implementation of a variant of Moore's algorithm to solve the single-source shortest-path problem illustrates two important ideas. First, contention for a single resource—software lockout—severely impairs the speedup achievable by their algorithm on a tightly coupled multiprocessor. Second, implementing halting conditions for relaxed algorithms is not always a straightforward task. Another single-source shortest-path algorithm, based on the linked array concept, eliminates the software lockout problem. Being a partitioned, rather than a relaxed, algorithm, the improved algorithm also has a cleaner halting condition.

BIBLIOGRAPHIC NOTES

A more detailed description of graph theoretic terms can be found in numerous texts [Harary 1969; Deo 1974; Reingold, Nievergelt, and Deo 1977]. Quinn and Deo [1984] have surveyed parallel graph algorithms.

Another parallel breadth-first SIMD search algorithm has been devised by Alton and Eckstein [1979]. (See also Eckstein [1979].) In addition, Eckstein and Alton [1977a, 1977b] have worked on the problem of parallel depth-first search with limited success. Recently Tiwari [1986] has explored the related problem of finding the depth-first spanning tree of a graph with a specified root, given the depth-first spanning tree of the same graph with a different root.

Most connected components algorithms in the literature use Hirschberg's approach of collapsing vertices. A number of graph problems are related to the connected components problem, including finding weakly connected components and strongly connected components in a digraph, finding lowest common ancestors, finding articulation points, finding biconnected and k-connected components, and planarity testing. References to connectivity-related algorithms in the literature include Attalah [1983]; Attalah and Kosaraju [1982]; Eckstein [1979a]; Guibas, Kung, and Thompson [1979]; Hambrusch [1982, 1983]; Kosaraju [1979]; Levialdi [1972]; Levitt and Kautz [1972]; Lipton and Valdes [1981]; Miller and Stout [1985a, 1986]; Nassimi and Sahni [1980b]; Nath and Maheshwari [1982]; Reghbati and Corneil [1978]; Reif [1982]; Reif and Spirakis [1982]; C. Savage [1977, 1981]; Savage and Ja'Ja' [1981]; Shiloach and Vishkin [1982a]; Tsin and Chin [1982]; van Scoy [1976]; and Wyllie [1979].

Browning [1980a, 1980b] has designed multicomputer algorithms that use an exponential number of processors to solve the maximum clique, color cost, and traveling salesperson problems.

Miller and Stout [1984a, 1984b] have proved that the SIMD-MC^2 model can find the convex hull of $O(n)$ planar points in $O(n^{1/2})$ time. They also present $O(n^{1/2})$ algorithms to decide whether two figures are linearly separable and to determine the nearest neighbor of each of $O(n)$ planar points.

Dekel and Sahni [1982] have developed an SIMD-SM algorithm to find the maximum matching of a convex bipartite graph.

Determining the minimum spanning forest of a graph can be seen as a simple variation of determining connected components. At each iteration the minimum edge, rather than the minimum labeled vertex, is found. Hence it is not surprising that Chin, Lam, and Chen [1982] have modified Hirschberg's algorithm to solve the minimum spanning tree problem by using the SIMD-SM-R model. Deo and Yoo [1981] have implemented parallelizations of Kruskal's, Sollin's, and Prim-Dijkstra's minimum spanning tree algorithms on the Denelcor HEP. They report that Cheriton and Tarjan's [1976] minimum spanning tree algorithm, when parallelized, is identical to the parallel version of Sollin's algorithm. Other references to parallel minimum spanning tree algorithms in the literature include Bentley [1980]; Atallah [1983]; Atallah and Kosaraju [1982, 1984]; Hambrusch [1982]; Hirschberg [1982]; Kučera [1982]; Levitt and Kautz [1972]; Reif [1982]; and C. Savage [1978].

Crane [1968] has proposed a parallel single-source shortest-path algorithm for use on an associative processor. Other shortest-path algorithms in the literature include those by Arjomandi [1975]; Deo, Pang, and Lord [1980]; Kučera [1982]; Levitt and Kautz [1972]; Mateti and Deo [1981]; Price [1982, 1983]; Quinn [1983]; and Yoo [1983].

Shiloach and Vishkin [1982b] have developed a parallel algorithm for finding the maximum flow in a directed, weighted graph. Their algorithm is based upon the usual layered network approach. Chen and Feng [1973] and Chen [1975] discuss parallel algorithms for the maximum-capacity path problem.

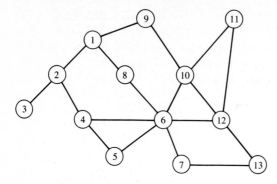

Figure 8-14 Undirected graph for Exercise 8-1.

Parallel algorithms for the stable-marriage problem have been discussed by Hull [1984], Quinn [1985], and Tseng and Lee [1984b].

EXERCISES

8-1 Illustrate the search tree resulting from p-depth search, parallel breadth-depth search, and parallel breadth-first search of the graph of Figure 8-14. Assume the following: $p = 3$, the search begins at vertex 1, and adjacent vertices are always explored in increasing order of the vertex numbers.

8-2 Prove Theorem 8-3.

8-3 Write an $\Theta(\log^2 n)$ transitive closure algorithm for the SIMD-SM-R model. Assume that $n = 2^k$, where k is a positive integer.

8-4 Analyze the complexity of Nassimi and Sahni's connected components algorithm.

8-5 Why is the number of trees in Sollin's algorithm reduced by at least a factor of 2 every iteration?

8-6 Give an example of a graph in which Sollin's algorithm requires only a single iteration to produce a minimum spanning tree.

8-7 Give an example of a graph in which Sollin's algorithm requires $\lceil \log n \rceil$ iterations to produce a minimum spanning tree.

8-8 Write a parallel version of Sollin's algorithm for the tightly coupled multiprocessor model.

8-9 Explain how Yoo's parallel heap emptying algorithm can be modified if $p < \log m$ processors are available.

8-10 In the parallel version of Moore's algorithm for the tightly coupled multiprocessor model, why is it not appropriate for a process to terminate when it discovers that the queue is empty?

8-11 Explain why the queue must be locked when process 1 examines the status of the array *waiting* in the tightly coupled multiprocessor version of Moore's algorithm.

8-12 Prove Theorem 8-13.

*8-13 Use of a linked array is one way to avoid the software lockout problem encountered by Deo, Pang, and Lord's parallel single-source shortest-path algorithm that used a single queue. Devise a parallel single-source shortest-path algorithm by using another data structure that avoids (or at least reduces) software lockout.

NINE

COMBINATORIAL SEARCH

Combinatorial algorithms perform computations on discrete, finite mathematical structures [Reingold, Nievergelt, and Deo 1977]. Combinatorial search is the process of finding "one or more optimal or suboptimal solutions in a defined problem space" [Wah, Li, and Yu 1985], and it has been used for such problems as laying out circuits in VLSI to minimize the area dedicated to wires, finding traveling salesperson tours, and theorem proving.

There are two kinds of combinatorial search problems. An algorithm to solve a **decision problem** must find a solution that satisfies all the constraints. An algorithm that solves an **optimization problem** must also minimize (or maximize) an objective function associated with solutions. All the examples of combinatorial search in this chapter are of optimization problems.

A search problem can be represented by a tree. The root of the tree represents the initial problem to be solved. The nonterminal nodes are either AND nodes or OR nodes. An AND node represents a problem or subproblem that is solved only when all its children have been solved; an OR node represents a problem or subproblem that is solved when any of its children has been solved. Every nonterminal node in an **AND tree** is an AND node (Figure 9-1a). The search tree corresponding to a divide-and-conquer algorithm is an AND tree, since the solution to a problem is found by combining the solutions to all its subproblems. Every nonterminal node in an **OR tree** is an OR node (Figure 9-1b). Branch-and-bound algorithms yield OR trees, since the solution to a problem can be found by solving any of its subproblems. An **AND/OR tree** is characterized

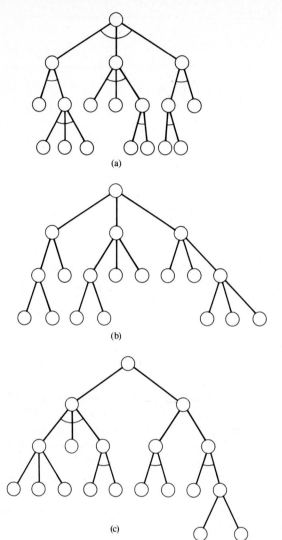

Figure 9-1 A search problem can be represented by a tree. (a) An AND tree. (b) An OR tree. (c) An AND/OR tree.

by the presence of both AND nonterminal nodes and OR nonterminal nodes (Figure 9-1c). Game trees are examples of AND/OR trees.

Section 9-1 discusses the parallel search of AND trees resulting from the application of divide-and-conquer algorithms. Section 9-2 discusses branch-and-bound algorithms in some detail. Proofs that branch-and-bound algorithms can exhibit anomalous speedup are presented. Increasing the number of processors can sometimes increase the execution time of certain searches. At other times increasing the number of processors can cause a disproportionate decrease in execution time. The alpha-beta search algorithm is presented in Section 9-3, and a number of methods to parallelize alpha-beta search are discussed.

9-1 DIVIDE AND CONQUER

Divide and conquer is a problem-solving methodology that involves partitioning a problem into subproblems, solving the subproblems, and then combining the solutions to the subproblems into a solution for the original problem. The methodology is recursive; that is, the subproblems themselves may be solved by the divide-and-conquer technique. The quicksort algorithm presented in Chapter 4 is an example of the divide-and-conquer technique.

The divide-and-conquer solution of a problem can be represented by an AND tree, since the solution to any problem represented by an interior node requires the solution of all its subproblems, represented by the children of that node. In other words, every node in the tree must be examined.

Three ways of executing divide-and-conquer algorithms on MIMD computers have been proposed. The first method is to build a tree of processors that corresponds to the search tree. This method has two disadvantages: The root processor can become a bottleneck, since it is the conduit for all input and output; and any particular processor tree, having a fixed interconnection structure, is appropriate for only some divide-and-conquer algorithms. A second method is to use a virtual tree machine that has a robust interconnection network, such as a multicomputer with a hypercubic processor organization. Since the virtual links between problems and subproblems may not be actualized by physical links, a good algorithm is needed to map subproblems to processors to minimize communication times. The third method suggests using tightly coupled multiprocessors to execute divide-and-conquer algorithms.

The parallel search of an AND tree can be divided into three phases. In the first phase problems are divided and propagated throughout the parallel computer. For most of the first phase there are fewer tasks than processors, and processors idle until they are given a problem to divide and propagate. In the second phase all the processors stay busy computing. In the third phase there are again fewer tasks than processors, and some processors combine results while other processors idle. Hence the maximum speedup achievable is limited by the propagation and combining overhead.

9-2 BRANCH AND BOUND

Backtrack is a familiar form of exhaustive search. The branch-and-bound method is a variant of backtrack that can take advantage of information about the optimality of partial solutions to avoid considering solutions that cannot be optimal. As an example of the branch-and-bound technique, consider the 8-puzzle (Figure 9-2), a simplified version of the 15-puzzle invented by Sam Loyd in 1878. The 8-puzzle consists of 8 tiles, numbered 1 through 8, arranged on a 3×3 board. Eight locations contain exactly one tile; the ninth location is empty. The object of the puzzle is to repeatedly fill the hole with a tile adjacent to it in the horizontal or vertical direction until the tiles are in row-major order.

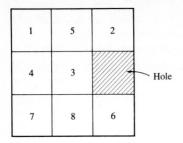

Figure 9-2 The 8-puzzle, a simplified version of the 15-puzzle invented by Sam Loyd in 1878.

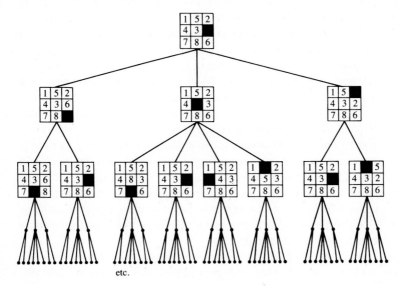

Figure 9-3 A portion of the state space tree corresponding to the search for a solution to a particular arrangement of the 8-puzzle.

Given an initial board position and a mechanism for generating legal moves from any position, it is possible to construct a tree of board positions that can be reached from the initial position. This tree is called the **state space tree** (Figure 9-3). One way to solve the puzzle is to pursue a breadth-first search of this state space tree until the sorted state is discovered. However, the goal is to examine as few alternative moves as possible. A means for achieving that goal is to associate with each state an estimate of the minimum number of tile moves needed to solve the puzzle, given the moves made so far.

One such function adds the number of tile moves made so far to the Manhattan distance between each out-of-place tile and its correct location. [The Manhattan distance between two tiles with (row, column) coordinates (x_1, y_1) and (x_2, y_2) is $|x_1 - y_1| + |x_2 - y_2|$.] Given such a function, the search can be concentrated on the portions of the state space tree that contain the most promising moves. At any time the search proceeds from the node having the smallest function value. If two nodes have the same value, then the node farther from the root of the state space tree is examined. If two nodes the same distance from the root have the same value, then one node is chosen arbitrarily. The branch-and-bound search of an 8-puzzle appears in Figure 9-4. Note that the algorithm requires that far fewer nodes be examined than in the breadth-first search.

Now that a concrete example has been given, we formally define the branch-and-bound technique. Given an initial problem and some objective function f to be minimized, a branch-and-bound algorithm attempts to solve it directly. If the problem is too large to be solved immediately, then it is decomposed into a set of two or more subproblems of smaller size. Every subproblem is characterized by the inclusion of one or more constraints. The decomposition process is repeated until each unexamined subproblem is decomposed, solved, or shown not to be leading to an optimal solution to the original problem.

In the 8-puzzle example, the problem is to put the pieces in order. The objective function f is the number of moves needed to order the pieces. If

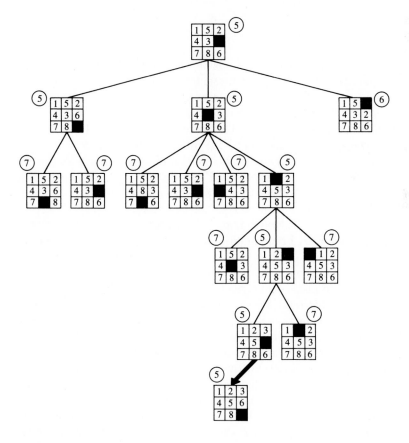

Figure 9-4 The branch-and-bound search for a solution to an arrangement of the 8-puzzle identical to that of Figure 9-3.

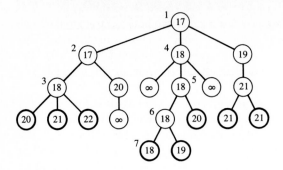

Figure 9-5 Another example of a state space tree. The values inside the nodes are the lower bounds of the solutions to the corresponding subproblems. Nodes corresponding to feasible solutions are represented by heavy circles. The best solution to this problem has cost 18. The numbers near the nodes represent the order in which they are examined by the best-first search strategy.

only a single move is necessary to order the pieces, the algorithm solves the problem directly. Otherwise, it decomposes the problem by generating a number of subproblems, one per legal move.

As we have seen in the case of the 8-puzzle, the decomposition process applied to the original problem may be represented by a rooted tree, called the **state space tree**. The nodes of this tree correspond to the decomposed problems, and the arcs of the tree correspond to the decomposition process. The original problem is the root of the tree. The leaves of the tree are those partial problems that are solved or discarded without further decomposition.

A branch-and-bound tree is distinguished in two important ways from trees representing divide-and-conquer algorithms. First, the tree is an **OR tree** (Figure 9-1b): the solution to any subproblem is a solution to the original problem. Hence the entire tree need not be searched. In fact, the state space tree representng a branch-and-bound algorithm may be infinite. This is the second important difference between branch-and-bound trees and divide-and-conquer trees.

Recall that the goal of the branch-and-bound technique is to solve the problem by examining a small number of elements in this tree. Assume that a minimum cost solution f^* is desired. A lower bounding function g is calculated for each decomposed subproblem as it is created. This lower bound represents the smallest possible cost of a solution to that subproblem, given the subproblem's constraints. On any path from the root to a terminal node, the lower bounds are always nondecreasing. In addition, the lower bound $g(x)$ at every leaf node x representing a feasible solution is identical to the value of the objective function $f(x)$ for that subproblem. Leaf nodes representing infeasible solutions have the value ∞. Figure 9-5 is another example of a state space tree. The values inside the nodes are the lower bounds of the corresponding subproblems. Nodes corresponding to feasible solutions are represented by heavy circles. The best solution to this problem has cost 18; i.e., the value of f^* is 18.

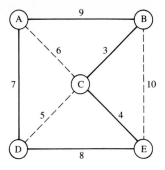

Figure 9-6 A weighted graph and its traveling salesperson tour.

At any point during the execution of a branch-and-bound algorithm there exists a set of problems that have been generated but not yet examined. A search strategy determines the order in which the unexamined subproblems are examined. The best-first (best-bound) search strategy selects the unexamined subproblem with the smallest lower bound. In the case of a tie, the subproblem deepest in the state space tree (i.e., the subproblem with the most constraints) is chosen. Ties unresolved by the deepness heuristic are broken arbitrarily. The numbers near the nodes in Figure 9-5 indicate the order in which the nodes are examined by the best-first search strategy.

A branch-and-bound algorithm can be characterized by how subproblems are generated, how a particular subproblem is selected as the point to continue the search, how hopeless subproblems are discarded, and how the algorithm terminates. Any of these steps can be performed in parallel.

Traveling Salesperson Problem

The traveling salesperson problem (TSP) is defined as follows: Given a set of vertices and a nonnegative cost $c_{i,j}$ associated with each pair of vertices i and j, find a circuit containing every vertex in the graph so that the cost of the entire tour is minimized. An example of a weighted graph and its traveling salesperson tour are given in Figure 9-6.

Little et al. [1963] devised a famous branch-and-bound algorithm to solve the traveling salesperson problem. When an unsolvable problem is encountered, it is broken into two subproblems representing tours that must include or exclude a particular edge. The edge that is to be used as the added constraint for the subproblems is chosen so that the lower bound on the cost of the solution of the subproblem excluding that edge is maximized. In other words, when a problem is broken into subproblems, the algorithm examines the minimum increase in the tour length when various edges are excluded and chooses the edge whose exclusion causes the largest increase in the tour length.

Reduction is used to find a lower bound on the cost of the tour, given the constraints made so far. The reduction algorithm works as follows: For every vertex i in the graph, the length c_i of the shortest edge leading to vertex i is found. If $c_i > 0$, then the lower bound can be increased by c_i if c_i is subtracted

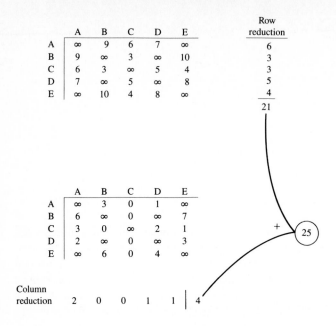

	A	B	C	D	E	Row reduction
A	∞	9	6	7	∞	6
B	9	∞	3	∞	10	3
C	6	3	∞	5	4	3
D	7	∞	5	∞	8	5
E	∞	10	4	8	∞	4
						21

	A	B	C	D	E
A	∞	3	0	1	∞
B	6	∞	0	∞	7
C	3	0	∞	2	1
D	2	∞	0	∞	3
E	∞	6	0	4	∞

Column reduction

| 2 | 0 | 0 | 1 | 1 | 4 |

	A	B	C	D	E
A	∞	3	0	0	∞
B	4	∞	0	∞	6
C	1	0	∞	1	0
D	0	∞	0	∞	2
E	∞	6	0	3	∞

Figure 9-7 An example of matrix reduction used by Little et al.'s traveling salesperson algorithm.

from the length of every edge leading to vertex i. After this step has been performed, the rows can be reduced in a similar fashion. For every vertex i in the graph, the length r_i of the shortest edge leading from vertex i is found. If $r_i > 0$, then the lower bound can be increased by r_i if r_i is subtracted from the length of every edge leading from vertex i. An example of matrix reduction appears in Figure 9-7.

A problem is broken into subproblems that are easier to solve because the subproblems contain additional constraints. In this algorithm, for example, including an edge reduces the number of edges that must be added to complete the tour; excluding an edge reduces the number of candidate edges. By driving up the lower bound as quickly as possible, the goal is to limit the number of subproblems (nodes in the state space tree) that must actually be examined. What

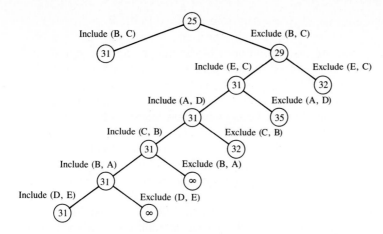

Figure 9-8 The state space tree corresponding to a best-first branch-and-bound search for a traveling salesperson tour of the graph shown in Figure 9-7.

follows is a high-level description of a best-first branch-and-bound algorithm using the best-first variant of Little et al.'s problem decomposition heuristic.

```
TRAVELING SALESPERSON (SISD):
begin
    reduce weight matrix, determining the root's lower bound
    initially only the root is in the state space tree
    {The root represents the set of all possible tours}
    repeat
        select the unexamined node in the state space tree
                with the smallest lower bound
        if the node represents a tour then exit the loop endif
        select the edge whose exclusion increases
                the lower bound the most
        for the two cases representing the inclusion and
                exclusion of the selected edge do
            create a child node with the correct constraint
            find the lower bound for the child node
        endfor
    forever
end
```

Figure 9-8 presents the state space tree corresponding to this algorithm's search for the traveling salesperson tour of the graph shown in Figure 9-7. Each node in the state space tree represents a set of possible tours satisfying constraints specified by the path from the root to that node. The root represents the set of all possible tours. Since reducing the weight matrix of the directed graph results in the value 25 being assigned to the root, a lower bound on the length of any tour is 25. The edge whose exclusion causes the greatest increase in the lower bound is (B,C). Hence the two children of the root represent the alternatives of including or excluding edge (B,C). Every tour explored in the left subtree must contain edge (B,C). No tour explored in the right subtree may contain edge (B,C). Given its constraint, each child node of the parent can be reduced, and lower bounds on all solutions based on that constraint can be determined. The lower bound of the left child is 31, while the lower bound on the right child is 29. In other words, tours that contain edge (B,C) must have length at least 31, while tours that do not contain edge (B,C) must have length at least 29. Hence the right child is the next node to be explored. Excluding edge (E,C) causes the greatest increase in the lower bound. The value in the right child—32—is a lower bound on all solutions that do not have edge (B,C) or edge (E,C). The value in the left child—31—is a lower bound on all solutions that do not have edge (B,C) but do have edge (E,C).

Since there are two unexplored nodes with the same lower bound (31), we explore the node deeper in the tree. This process continues until a tour is discovered, at which point the algorithm terminates. The tour contains edges (E,C), (A,D), (C,B), (B,A), and (D,E).

Two Parallel Traveling Salesperson Algorithms. Mohan [1983] has developed two parallelizations of this algorithm. The first parallel algorithm involves a parallelization of the **for** loop; the second parallel algorithm executes the **repeat** loop in parallel.

As presented before, the **for** loop has a natural parallelism of 2—each node has only two children. However, by selecting k edges to be considered for inclusion or exclusion, the number of children of each node increases to 2^k, since constraints reflecting all combinations of inclusion and exclusion must be generated. The modified algorithm follows. Clearly this partitioned algorithm is appropriate for 2^k processors.

```
TRAVELING SALESPERSON (TIGHTLY COUPLED MULTIPROCESSOR):
begin
    reduce weight matrix, determining the root's lower bound
    initially only the root is in the state space tree
    repeat
        select the unexamined node in the state space tree
                with the smallest lower bound
        if the node represents a tour then exit the loop endif
        select the k edges whose exclusion increases
                the lower bound the most
        for the 2^k cases representing all inclusion-exclusion
                combinations of the selected edges do
            create a child node with the correct constraints
            find the lower bound for the child node
        endfor
    forever
end
```

The second algorithm creates a number of processes that asynchronously explore the tree of subproblems until a solution has been found. Each process

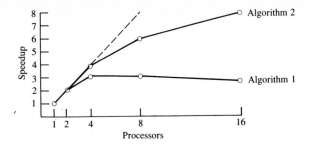

Figure 9-9 The speedup of Mohan's two parallel algorithms to solve the traveling salesperson problem with 30 vertices on Cm*.

repeatedly removes the unexplored subproblem with the smallest lower bound from the ordered list of unexplored subproblems, decomposes the problem (unless it can be solved directly), and inserts the two newly created subproblems in their proper places in the ordered list of problems to be examined. A process must have exclusive control of the list in order to insert and delete elements, but the time taken for these tasks is relatively small compared to the time needed to decompose a problem. Thus contention for this list should not be a significant inhibitor of speedup.

The speedup of these two parallel algorithms on Cm* is contrasted in Figure 9-9. The first algorithm achieves extremely poor speedup. The additional processors spend most of their time creating nodes that are never explored, because their lower bounds are too high. Mohan's second algorithm achieves a speedup of about 8 with 16 processors when solving a 30-vertex TSP. The major obstacle to higher speedup is intracluster contention, or contention by computer modules within a cluster for shared resources. These resources include the Kmap, the Map bus, and the Object Manager. The Object Manager is used to create the nodes of the search tree.

Anomalies in Branch and Bound

Mohan has shown that a parallelization of the best-first branch-and-bound technique that allows more than one node to be examined at a time can achieve good speedup. In this section we present Lai and Sahni's [1983] analysis of the speedups theoretically achievable by such an algorithm. A few assumptions must be made in order for the analysis to be manageable. Assume that the time needed to examine any node in the tree and decompose it is constant for all nodes in the state space tree. Furthermore, assume that execution of the parallel algorithm consists of a number of "iterations." During each iteration every processor examines a unique subproblem, if one is available, and decomposes it. Given a particular branch-and-bound problem to be solved and a particular lower bounding function g, define $I(p)$ to be the number of iterations required to find a solution node when p processors are used.

The first theorem shows that increasing the number of processors can actually increase the number of iterations required to find a solution.

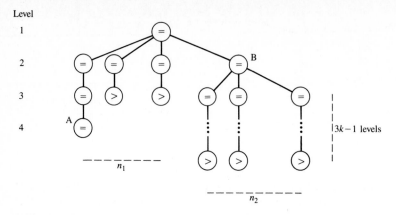

Figure 9-10 A state space tree illustrating that increasing the number of processors may actually increase the execution time of a branch-and-bound algorithm.

Theorem 9-1 (See Lai and Sahni [1983].) Given $n_1 < n_2$ and $k > 0$, there exists a state space tree such that $kI(n_1) < I(n_2)$.

Proof. Consider the state space tree shown in Figure 9-10. All nodes labeled "=" have the same lower bound, which happens to be the value of the least-cost answer node (node A). Nodes labeled ">" have a lower bound greater than the value of the least-cost answer node. When n_1 processors conduct the search, on the first iteration the root node is expanded into $n_1 + 1$ children nodes. The second iteration consists of expanding the n_1 leftmost nodes at level 2 into n_1 nodes at level 3. Of the nodes at level 3, $n_1 - 1$ of them cannot lead to the solution and are discarded. On iteration 3 the remaining node at level 3 and node B are expanded. Since the node at level 3 leads to the solution node, the algorithm terminates. Hence $I(n_1) = 3$.

When n_2 processors conduct the search, the first iteration is the same: The root node is expanded into $n_1 + 1$ children nodes. On the second iteration, however, all $n_1 + 1$ nodes at level 2 are expanded, yielding $n_1 + n_2$ nodes at level 3. Since only n_2 nodes at level 3 can be expanded on iteration 3, it could happen that the n_2 rightmost nodes would be the ones chosen. If we assume the processors expanded the n_2 rightmost nodes at level 3, n_2 nodes at level 4 would be created, and iterations 4, 5, 6, $\ldots$, $3k$ could be devoted to a wild goose chase, expanding nodes down the right part of the tree. The solution node A would be expanded on iteration $3k + 1$. Hence $I(n_2) = 3k + 1$. Combining the two results yields $kI(n_1) = 3k < 3k + 1 < I(n_2)$. ∎

The fact that a large number of nodes have a lower bound equal to the value of the least-cost answer node—f^*—leads to the anomaly described in the previous theorem. What would happen if $g(x) \neq f^*$ whenever x is not a solution node?

Definition 9-1 A node x is **critical** if $g(x) < f^*$.

Theorem 9-2 (See Lai and Sahni [1983].) If $g(x) \neq f^*$ whenever x is not a solution node, then $I(1) \geq I(n)$ for all $n > 1$.

Proof. By the definition of the best-first branch-and-bound heuristic, only critical nodes and least-cost answer nodes can be expanded. In addition, every critical node must be expanded before any least-cost answer node is expanded. Hence if the number of critical nodes is m, then $I(1) = m$. When $n > 1$, at least one of the nodes expanded each iteration must be a critical node. (Why?) Hence a least-cost answer node must be examined no later than iteration m. Thus if the number of critical nodes is m, then $I(n) \leq m$. Therefore $I(1) \geq I(n)$ for all $n > 1$. ∎

The following theorem proves that increasing the number of processors can actually cause a disproportionate decrease in the number of iterations required to find a solution node.

Theorem 9-3 (See Lai and Sahni [1983].) Given $n_1 < n_2$ and $k > n_2/n_1$, then there exists a state space tree such that $I(n_1)/I(n_2) \geq k > n_2/n_1$.

Proof. This is left to the reader. ∎

Theorem 9-4 (See Lai and Sahni [1983].) If $g(x) \neq f^*$ whenever x is not a least-cost answer node, then $I(1)/I(n) \leq n$ for $n > 1$.

Proof. Let m be the number of critical nodes. Then $I(1) = m$ (Theorem 9-2). All critical nodes must be expanded before the parallel branch-and-bound algorithm can terminate. (Why?) Hence $I(n) \geq m/n$, or $I(1)/I(n) \leq n$. ∎

Lai and Sahni have found anomalous behavior in some instances of the 0-1 knapsack problem, but they conclude that anomalous behavior is rarely encountered in practice and that in general (1) increasing the number of processors will not increase execution time (assuming the problem is large enough) and (2) superlinear speedup cannot be expected.

9-3 ALPHA-BETA SEARCH

The most successful computer programs to play two-person zero-sum games of perfect information, such as chess, checkers, and go, have been based on exhaustive search algorithms. These algorithms consider series of possible moves and countermoves, evaluate the desirability of the resulting board positions, then work their way back up the tree of moves to determine the best initial move.

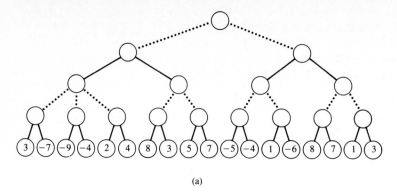

(a)

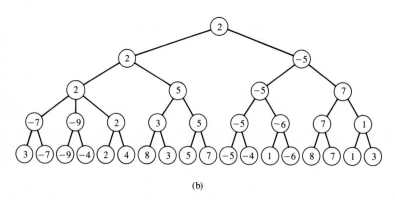

(b)

Figure 9-11 (a) A game tree. Dotted edges represent the moves available to the first player; solid edges represent moves available to the second player. (b) The same tree with the values of the interior nodes filled in.

Given a trivial game, the **minimax** algorithm can be used to determine the best strategy. Figure 9-11a represents the **game tree** of a hypothetical game played for money, whose rules are left unstated. Dotted edges represent moves made by the first player; solid lines represent moves made by the second player. The root of the tree is the initial condition of the game. The leaves of this game tree represent outcomes of the game. Interior nodes represent intermediate conditions. The outcomes are always put in terms of advantage to the first player. Thus positive numbers indicate the amount of money won by the first player, while negative numbers indicate the amount of money lost by the first player. The algorithm assumes that the second player tries to minimize the gain of the first player, while the first player tries to maximize his or her own gain, hence the name of the algorithm. Figure 9-11b is the same tree with the values of the interior nodes filled in. The value of this game to the first player is 2. If the first player plays the minimax strategy, she or he is guaranteed to win at least that much.

Stockman [1979] has pointed out that a game tree is an example of an AND/OR tree. The AND nodes represent positions resulting from a move by MIN—the root and all nodes an even distance from the root. The OR nodes represent positions resulting from a move by MAX—all nodes an odd distance from the root.

Interesting games such as chess have game trees that are far too complicated to be evaluated exactly. For example, de Groot has estimated that there may be 38^{84} positions in a chess game tree [de Groot 1965]. Thus current chess-playing programs examine moves and countermoves only to a certain depth, then estimate the value of the board position to the first player at that point. Of course, evaluation functions are unreliable. If a perfect evaluation function existed, the need for searching would be eliminated. (Why?) As we have seen, all possible moves and countermoves from a position p to some predetermined lookahead horizon can be represented by a game tree. The minimax value of the game tree can be found by applying the evaluation function to the leaves of the tree (the terminal nodes), then working backward up the tree. If it is the second player's move at a particular nonterminal node in the game tree, the value assigned is the minimum over all its children nodes. If it is the first player's move, the value assigned is the maximum over all its children nodes. The minimax algorithm examines every terminal and nonterminal node in the tree.

It is generally true that the deeper the search, the better the quality of play. That is why **alpha-beta pruning** has proved itself so valuable. Alpha-beta pruning, a form of branch-and-bound algorithm, avoids searching subtrees whose evaluation cannot influence the outcome of the search, i.e., cannot change the choice of move. Hence it allows a deeper search in the same amount of time.

The alpha-beta algorithm is called with four arguments: p, the current condition of the game; *alpha* and *beta*, the range of values over which the search is to be made; and *depth*, the depth of the search that is to be made. The function returns the minimax value of the position p. What follows is Knuth's "negamax" version of the alpha-beta algorithm [Marsland and Campbell 1982]. It has the advantage that minimizing and maximizing occur correctly without explicit tests for whose move it is.

```
function ALPHA_BETA (p, alpha, beta, depth)
begin
    if depth ≤ 0 then
        return (EVALUATE (p))        {Evaluate terminal node}
    else
        width ← GENERATE(p)          {width is number of legal moves}
                                     {Children of p are p.1...p.width}
        if width = 0 then
            return (EVALUATE (p))    {No legal moves}
```

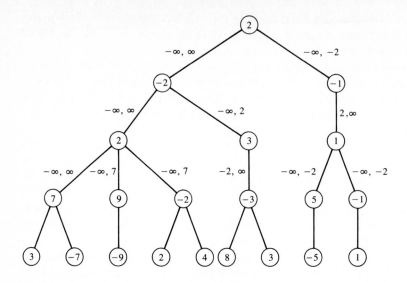

Figure 9-12 An illustration of alpha-beta search.

```
        else
          score ← alpha
          for i ← 1 to width do
            MAKE (p.i)                {Make the move}
            value ← −ALPHA_BETA (p.i, −beta, −score, depth − 1)
            UNDO (p.i)               {Unmake the move}
            if value > score then
              score ← value          {A better move has been found}
            endif
            if score ≥ beta then
              return (score)         {A cutoff has been found}
            endif
          endfor
        endif
      endif
    return (score)
  end
```

To illustrate the workings of Knuth's algorithm, consider the game tree in Figure 9-12. This tree represents the same game as the tree shown to illustrate the minimax algorithm, except that nodes not examined by the alpha-beta algorithm have not been drawn.

The pair of values on each arc represents the values of *alpha* and *beta* passed to the lower node in the arc. Look at the leaves of the search tree. The values of *alpha* and *beta* passed to these nodes are not shown, since they are irrelevant. Because *depth* ≤ 0, the evaluation function is used to determine the value of that position, and this value is returned immediately to the parent node.

Now consider the interior nodes. Each represents a position in the game.

When the search reaches one of these nodes, we know that some choice of moves that has already been considered leads to a value of at least *alpha* for the player whose move it is. We also know that there is no way that the opponent will let us get a value more than *beta* from the current position. Hence *alpha* and *beta* define a window for the search. If no moves from this point result in a value greater than *alpha*, then the player will not follow this course of action. If any moves from this point result in a value greater than *beta*, then the player's opponent will make sure that this position is never reached. Hence the negamax algorithm finds the maximum of the negative of the values returned by the children nodes; but if the negative of any value returned is greater than *beta*, then the search is terminated, and the rest of the alternatives are pruned.

Note that one effect of the negamax algorithm is to present the value in each node from the point of view of the person making the move. Hence the root and all nodes an even distance from the root contain values representing the worth of the position to the player making the first move, while all nodes an odd distance from the root contain values representing the worth of the position to the player making the second move. Finally, note that the negamax version of the alpha-beta algorithm finds the same value and selects the same line of play as the minimax algorithm, while sharply reducing the number of nodes examined.

The analysis of the complexity of the alpha-beta algorithm is simplified by assuming, as in Marsland and Campbell [1982], that the trees to be searched are uniform—that is, that all nonterminals have W children and that all terminal nodes are D levels deep in the search tree. The minimax algorithm searches every terminal node; hence

$$M(W, D) = W^D$$

Slagle and Dixon [1969] have shown that in the best case the alpha-beta algorithm examines many fewer nodes:

$$B(W, D) = W^{\lceil D/2 \rceil} + W^{\lfloor D/2 \rfloor} - 1$$

In other words, in the best case it is possible for the alpha-beta algorithm to examine approximately 2 times the square root of the number of nodes searched by the minimax algorithm. However, the best case assumes that the first move considered from any position is always the best move.

Two other kinds of search trees are of interest. A **random uniform** game tree assumes that the values of the terminal nodes are randomly chosen from some uniform distribution. Let $R(W, D)$ be the average number of terminal nodes evaluated by the alpha-beta algorithm when searching a random uniform game tree. A **strongly ordered** game tree assumes that (1) 70 percent of the time the first move chosen from any nonterminal node is the best move, and (2) 90 percent of the time the best move from any nonterminal node is one of the first 25 percent of the moves searched. These assumptions are not arbitrary, and strongly ordered game trees can be generated for real problems [Gillogly 1978; Marsland and Rushton 1974]. Let $S(W, D)$ denote the average number of terminal nodes evaluated by the alpha-beta algorithm when searching a strongly ordered game tree. Experiments indicate that

$$B(W, D) < S(W, D) < R(W, D) \ll M(W, D) = W^D$$

Hence under realistic conditions the alpha-beta algorithm performs much better than minimax.

The standard alpha-beta algorithm can be improved in a number of ways, further reducing the number of nodes to be examined. The following techniques have all proved themselves in practice.

One improvement is called **aspiration search**. Given a tighter "window" of values to search, the alpha-beta algorithm can prune many more branches and complete the search much more quickly. Aspiration search works by making an estimate of the value v of the board position at the root of the game tree, figuring the probable error e of that estimate, then calling the alpha-beta algorithm with the initial window $(v - e, v + e)$. If the value of the game tree does indeed fall within this window of values, then the search will complete sooner than if the algorithm had been called with the initial window $(-\infty, \infty)$. If the value of the game tree is less than $v - e$, then the search will return the value $v - e$, and the algorithm must be called again with another window, such as $(-\infty, v - e)$. Similarly, if the value of the game tree is greater than $v + e$, then the search returns the value $v + e$, and another search will have to be done with a modified initial window, such as $(v + e, \infty)$.

In many two-person games it is common for the same position to appear at many different places in the game tree. An alternative to evaluating the same position repeatedly is to store previously evaluated positions in a large hash table, called a **transposition table**. Before a position is evaluated, the transposition table is checked to see whether a value for that position has already been found. Because of the artificiality of game trees, nearly perfect hashing functions can be developed, so that this check can be done quickly.

Any move that causes a cutoff in the search tree at level i is said to **refute** the move at level $i - 1$ [Cichelli 1973]. The **killer heuristic** applies such moves early in the search at all positions at level i, hoping to cause quick cutoffs at these positions, too. The killer heuristic has an additional benefit. Since it causes the same set of moves to be considered over and over, it increases the usefulness of the transposition table.

Another variant on the standard alpha-beta algorithm is called **iterative deepening**. Each level of a game tree is called a **ply** and corresponds to the moves of one of the players. Iterative deepening is the use of a $(D-1)$-ply search to prepare for a D-ply search. This technique has three advantages [Marsland and Campbell 1982]. First, it allows the time spent in a search to be controlled. The search can continue deeper and deeper into the game tree until the time allotted has expired. Second, a $(D - 1)$-ply search can provide a principal variation for the D-ply search, allowing the alpha-beta search to execute more quickly. Finally, the value returned from a $(D - 1)$-ply search can be used as the center of the window for a D-ply aspiration search.

Parallel Alpha-Beta Search

Alpha-beta search has a number of opportunities for parallel execution. The simplest method is to parallelize the evaluation of a board position. A large percentage of the time spent by a program performing an alpha-beta search is dedicated to evaluating terminal positions. For example, about 40 percent of the time spent by the chess-playing programs BLITZ and DUCHESS is used in evaluating board positions. By assigning different processors to different terms of the scoring function, the process could be speeded up. Marsland and Campbell note three advantages of this technique. First, the search can be made deeper, since the evaluation time is reduced. Second, a large number of processors could be used to evaluate particular features of a game position. Third, more complicated evaluation functions could be devised, limited only by the number of processors available.

Another straightforward parallelization of the alpha-beta algorithm is to perform aspiration search in parallel. If three processors were available, then each processor could be assigned one of the windows $(-\infty, w-e)$, $(w-e, w+e)$, and $(w+e, \infty)$. Ideally the processor searching $(w-e, w+e)$ will succeed, but all three processors will finish before a single processor searching the window $(-\infty, \infty)$. Baudet [1978a, 1978b] explored parallel aspiration on Cm*.

Work on parallel aspiration has led to two conclusions. First, the maximum expected speedup is typically 5 or 6, regardless of the number of available processors. This is because $B(W, D)$ is a lower bound on the cost of aspiration search. Second, parallel aspiration search can sometimes lead to superlinear speedup when two or three processors are being used.

Parallel aspiration search seems to be more promising when aspiration search is used in conjunction with iterative deepening, since one of the problems with sequential algorithms using aspiration search is that their initial windows tend to be too wide.

There are a number of ways to parallelize the use of the transposition table. One method, suitable for two processors, is for one processor to perform the sequential alpha-beta search algorithm while the second processor manages the transposition table. When the first processor encounters a position, it asks the second processor for the value of that position. While the second processor looks for the position in the transposition table, the first processor can begin to evaluate it itself. If the position is not in the table, then no time is lost searching for the position. If the position is in the table, then the time needed to come up with a value is the minimum of the time spent by processor 1 evaluating the position and the time spent by processor 2 retrieving the value.

Of course, this method could be extended to situations in which there are a large number of processors searching the game tree. If there are more requests to the transposition table manager than a single processor can handle, then the task of managing the transposition table (performing insertions and searches) can be divided among a number of processors.

Many believe that significant speedups can only be achieved by allowing processors to examine independent subtrees in parallel. There are two important overheads to be considered. *Search overhead* refers to the increase in the number of nodes that must be examined owing to the introduction of parallelism. *Communication overhead* refers to the time spent coordinating the processes performing the searching. Search overhead can be reduced at the expense of communication overhead by keeping every processor aware of the current search window. Communication overhead can be reduced at the expense of search overhead by allowing processors to work with outdated search windows.

For example, consider this simple method of performing alpha-beta search in parallel. Split the game tree at the root, and give every processor an equal share of the subtrees. Let every processor perform an alpha-beta search on its subtrees. Each processor begins with the search window $(-\infty, \infty)$, and no processor ever notifies other processors of the changes in its search window. Clearly this algorithm minimizes communication overhead. What is the search overhead of this algorithm?

Assume that the processors are searching a game tree with $W = 40$ (a reasonable value for chess, for example). Sequential alpha-beta search of a tree with $W = 40$ is equivalent to an exhaustive search of a tree with $W = 7$ [Gillogly 1972]. If 40 processors were applied to the root of the game tree, then the average speedup achieved would be 7. Hence the amount of search overhead is significant: only $\frac{7}{40}$ of the nodes searched by each processor in the parallel algorithm are considered by the sequential algorithm.

9-4 SUMMARY

One way to differentiate between combinatorial search problems is to categorize them by the kind of state space tree they traverse. Divide-and-conquer algorithms traverse AND trees: the solution to a problem or subproblem has been found only when the solution to all its children has been found. Branch-and-bound algorithms traverse OR trees: the solution to a problem or subproblem can be found by solving any of its children. Game trees contain both AND nonterminal nodes and OR nonterminal nodes.

Parallel combinatorial search algorithms for all these trees have been proposed. The speedup achievable through the parallel search of an AND tree is limited by propagation and combining overhead.

Mohan's solution of the traveling salesperson problem on Cm* demonstrates the potential for implementing branch-and-bound algorithms on MIMD computers. The parallel algorithm achieves good speedup by good load balancing and the large granularity of subproblems. The fundamental problem faced by designers of parallel branch-and-bound algorithms is to keep the efficiency of the processors high by focusing the search on nodes that the sequential algorithm examines.

Lai and Sahni have given examples of state space trees for which parallel best-first branch-and-bound algorithms can show anomalous behavior, such as superlinear speedup. Experiments they have performed with the simulated parallel solution of the 0-1 knapsack problem show that anomalous behavior can occur in the "real world," albeit rarely.

Alpha-beta search has shown itself to be an efficient method for evaluating game trees. A number of improvements on standard alpha-beta search have been invented, including aspiration search, iterative deepening, and transposition tables. These methods are amenable to parallelization. To significantly improve the speed of alpha-beta search, however, independent subtrees must be searched in parallel. Minimizing communication overhead can cause an unacceptable amount of search overhead, and vice versa. It remains to be seen how great a speedup in alpha-beta search can be achieved in practice.

BIBLIOGRAPHIC NOTES

Ibaraki [1976a, 1976b] has written important papers analyzing branch-and-bound algorithms. Imai, Fukumara, and Yoshida [1979] consider a parallel branch-and-bound algorithm. Wah, Li, and Yu [1984, 1985] discuss Manip, a computer specifically designed to execute best-first branch-and-bound algorithms. They also describe parallelism in dynamic programming. Li and Wah [1984a, 1984b, 1985] have written a number of related papers. Kumar and Kanal [1983] recast branch-and-bound algorithms to encompass AND/OR tree search as well as OR tree search. In a later paper Kumar and Kanal [1984] discuss the parallelization of this branch-and-bound algorithm.

Deriving results similar to those of Lai and Sahni [1983], Quinn and Deo [1983b, 1986] describe anomalous behavior of branch-and-bound algorithms.

Quinn and Deo [1983a] have implemented a parallelization of the farthest insertion heuristic for the traveling salesperson problem on the Denelcor HEP. Their algorithm finds a nonoptimal solution in polynomial time. The farthest insertion heuristic and other approximate algorithms for the traveling salesperson problem are examined in Golden et al. [1980] and Rosenkrantz, Stearns, and Lewis [1974].

Marsland and Campbell [1982] have written an excellent, very readable survey paper describing the parallel search of strongly ordered game trees. Their treatment is much more detailed than the one presented in this chapter, and they cover many more interesting variations of parallel alpha-beta search. Parallel alpha-beta search has also been explored by Finkel and Fishburn [1982]. Knuth and Moore [1975] analyzed the sequential alpha-beta algorithm.

Earlier in the chapter we pointed out that the quality of play usually improves as the game tree search is deepened. Nau has shown that there exists an infinite class of pathological game trees for which this rule of thumb is not true. However, he adds that "pathology does not occur in games such as chess or checkers" [Nau 1982].

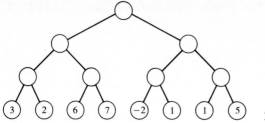

Figure 9-13

EXERCISES

9-1 Explain this statement by Wah, Li, and Yu [1985]: "Multiprocessing is generally used to improve the computational efficiency of solving a given problem, *not to extend the solvable problem space of the problem*" (their italics).

9-2 Given a divide-and-conquer algorithm whose complexity is described by the recurrence relation

$$T(n) = \Omega(n) + kT\left(\frac{n}{k}\right)$$
$$T(1) = \Omega(1)$$

prove a lower bound on the complexity of a parallel divide-and-conquer algorithm, assuming that the decomposition step [with complexity $\Omega(n)$] cannot be made parallel.

9-3 Using the 8-puzzle as an example, explain why it is likely that a depth-first search of the state space tree would not yield a solution.

9-4 Justify the deepness heuristic used by the best-first search strategy to choose between unexamined subproblems sharing the smallest lower bound.

9-5 In the context of Theorem 9-2, why must at least one critical node be examined every iteration when $p > 1$?

9-6 Prove Theorem 9-3.

9-7 In the context of Theorem 9-4, why must all critical nodes be expanded before the algorithm can terminate?

9-8 Explain the difficulties faced by an implementor of a parallel best-first branch-and-bound algorithm on a multicomputer. Describe possible solutions to these difficulties.

9-9 "If a perfect evaluation function existed, the need for searching would be eliminated." Explain.

9-10 Use the minimax algorithm to evaluate the game tree of Figure 9-13.

9-11 Use the alpha-beta algorithm to evaluate the game tree of Figure 9-13.

9-12 List methods used to parallelize alpha-beta search.

TEN

LOGIC PROGRAMMING

A new area of research into parallel algorithms, exemplified by the Japanese fifth-generation computer project, is the attempt to build systems capable of knowledge processing. Section 10-1 discusses logic programming, a way to implement knowledge processing systems. Prolog, a logic programming language, is briefly described in Section 10-2. Prolog seems amenable to several kinds of parallelism: OR parallelism, AND parallelism, stream parallelism, and parallel unification. Section 10-3 presents Concurrent Prolog, a parallel version of Prolog. Finally, Section 10-4 discusses the Bagel, an experimental parallel computer whose kernel language is Concurrent Prolog.

10-1 LOGIC PROGRAMMING

Logic investigates the relationships between premises and conclusions of arguments. To prove that a conclusion follows from a set of premises, a number of inference steps must be given, showing how the premises lead to the conclusion by following certain rules. Since natural language is ambiguous and complex, a symbolic language is used to express the premises and tentative conclusions. The goal of simplicity is further achieved by using the clausal form of logic which, though simple, is quite powerful.

Atomic sentences, the simplest sentences in clausal logic, express relationships:

1. Abby likes Prolog.
2. Prolog is fast.

Another kind of sentence has an atomic condition implying an atomic conclusion:

3. Brian likes Prolog if Prolog is portable.

It is also possible for a sentence to have multiple conditions:

4. Cathy uses Prolog if Prolog is fast and Prolog is available.

Sentences may have multiple conclusions as well:

5. Dennis is rich or Dennis is unlucky if Dennis sells Prolog.

A **Horn clause** is a clause that contains at most one conclusion. Sentences 1 to 4 can be represented as Horn clauses; sentence 5 cannot. **A logic program** is a set of Horn clauses, or **procedures**. Execution of a logic program begins when it is given a goal clause. The system tries to determine whether the goal clause is **satisfiable** (true under the rules of logic), given the logic program (facts about relationships). Prolog, designed and implemented in 1972, was the first logic programming system [Colmerauer et al. 1973, Roussel 1975]. It is based on the procedural implementation of Horn clauses.

Because the fundamental operation during the execution of a logic program is the application of a Horn clause, or procedure, the performance of logic programming systems is measured in terms of procedures invoked per second, more generally called **logical inferences per second** (LIPS).

Why was logic programming chosen as the mechanism for implementing fifth-generation computers? According to Fuchi, several trends in computer science point toward logic programming [Shapiro 1983a]:

Work in software engineering is making clear the importance of good specifications and specification languages. Symbolic logic is a natural vehicle for specifying a system. It is easier to specify and verify a system if the specification and the programming language use the same formalism.

Relational data bases have become more and more popular, partly because they are easy for a user to understand. Logic programs can be seen as a generalization of relational data bases.

Logic programming provides a natural mechanism for designing a single-assignment language. Hence programs written in such a language may be more easily implemented on a parallel computer.

Expert systems have two components: a knowledge base and an inference engine. The rule base fits naturally into Prolog. The deduction mechanism in an expert system is similar to Prolog's deduction mechanism.

10-2 PROLOG

Prolog is a programming language especially well suited to symbolic processing. The purpose of this section is to give you some feeling for what a Prolog program looks like and how it is executed.

Consider the following Prolog program (the numbers are for reference purposes only):

```
1.  male(david).
2.  male(edward).
3.  parent(david,thomas).
4.  parent(david,stuart).
5.  parent(edward,mary).
6.  parent(edward,valerie).
7.  male(thomas).
8.  male(stuart).
9.  female(mary).
10. female(valerie).
11. married(stuart,mary).
12. married(mary,stuart).
13. parent(thomas,loretta).
14. parent(valerie,kathleen).
15. female(loretta).
16. female(kathleen).
17. sibling(X,Y) :- parent(Z,X), parent(Z,Y), (X\==Y).
18. in_law(X,Y) :- married(X,Z), sibling(Z,Y).
19. aunt(X,Y) :- female(X), sibling(X,Z), parent(Z,Y).
20. aunt(X,Y) :- female(X), in_law(X,Z), parent(Z,Y).
```

The first 16 lines of this program represent one kind of statement; the last four lines represent another. Note that the format of these statements has been changed slightly from the discussion in the previous section. In Prolog the relationship appears first, followed by the object(s) to whom the relationship applies. Relationships and objects are denoted by a string of letters, digits, and some special characters, beginning with a lowercase letter. The first group of 16 statements has a head but no body. These represent **facts**. Line 1 can be understood to mean "David is a male," while line 5 can be understood to mean "Edward is the parent of Mary," and line 11 can be interpreted "Stuart is married to Mary." The last group of four lines are called **rules**. The capitalized letters in the rules are called **variables**. Variables stand for objects whose identities are unspecified. In this program, variables stand for arbitrary people.

Rules have both a declarative and a procedural interpretation. For example, a declarative interpretation of the first rule is "Person X is a sibling of person Y if there is a person Z such that (1) Z is a parent of X, (2) Z is a parent of Y, and (3) X and Y are distinct persons." A procedural interpretation of the first rule is "In order to prove that person X is a sibling of person Y, you must find a person Z and prove three things: (1) Z is a parent of X, (2) Z is a parent of Y, and (3) X and Y are distinct persons."

Note that there are two rules defining the aunt relationship. Person X is an aunt of person Y if X is female and there exists a third person Z such that X is an in-law of Z and Z is a parent of Y. It is also true that person X is an aunt of person Y if X is female and there exists a third person Z such that X is a sibling of Z and Z is a parent of Y.

Now that we have a set of facts about a family tree and rules defining certain familial relationships, we can ask the Prolog system to perform a computation. A computation of a Prolog program begins with some initial goal A and produces two kinds of results. First, the computation produces the answer yes or no to indicate whether the goal is satisfiable or not. Second, the final values of the variables in A that were originally unspecified may be considered the output of the computation. For example, giving the Prolog system the goal

aunt(mary,kathleen).

essentially asks the question "Is Mary the aunt of Kathleen?" The system will respond yes or no, depending on the facts and rules it has to work with. However, giving the Prolog system the goal

aunt(mary,X).

asks the question "Does Mary have any nieces or nephews?" If Mary does not, then the system will respond no. Otherwise, the system will provide a value for X that satisfies the goal.

Let us see how Prolog would handle the first query—whether Mary is the aunt of Kathleen. The goal statement is not explicitly listed as a fact. However, there are two rules governing "aunthood," and either may be appropriate in this circumstance. According to the definition of Prolog, the rule appearing first in the program is the first rule tried. That rule is

aunt(X,Y) :- female(X), in_law(X,Z), parent(Z,Y).

Now the single goal has been broken up into three subgoals, each of which must be true for the original goal to be satisfied. The process of breaking goals up into subgoals can be described by a search tree. Figure 10-1 is a drawing of the search tree corresponding to this question. The object of this particular search is to determine whether the root has the value *true* or *false*. This tree is an AND/OR tree, like those described in the previous chapter. Conjunctions are marked with arcs under the parent node. Disjunctions have no special markings. For example, aunt(mary,kathleen) is *true* if the values of either of the two subtrees is *true*. On the other hand, sibling(mary,X) is *true* if and only if an object can be found for variables X and Y such that parent(Y,mary) is *true* and parent(Y,X)

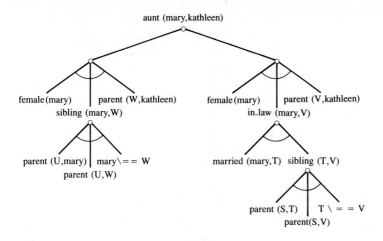

Figure 10-1 A search tree corresponding to the goal aunt(mary,kathleen).

is *true*. The Prolog system performs a depth-first search of this tree.

Figure 10-2 shows how a Prolog system would perform the search. The system works down the list of clauses, looking for a clause whose head matches the goal statement

aunt(mary,kathleen).

The first clause matching this goal is

aunt(X,Y) :- female(X), sibling(X,Z), parent(Z,Y).

The head of the clause is **unified** with the goal; that is, the variable X is instantiated to mary and the variable Y is instantiated to kathleen. Unification serves several purposes, including parameter passing and assigning values to variables. The clause is a rule; its head is *true* if its body is *true*. What we have done is reduce the goal aunt(mary,kathleen) into the conjunction of three goals. Hence the system must now determine if the body is *true*.

The latest derived clause is

female(mary), sibling(mary,Z), parent(Z,kathleen).

Prolog always attempts to match the first literal in the latest derived goal clause. In addition, the new goals derived from the matched literal are placed at the front of the goal clause. Hence the search proceeds in a depth-first manner.

The system now tries to satisfy the goal female(mary). Searching through the program, Prolog finds a clause that matches the goal. Since the clause has no body, nothing replaces the literal. Hence the latest goal clause is

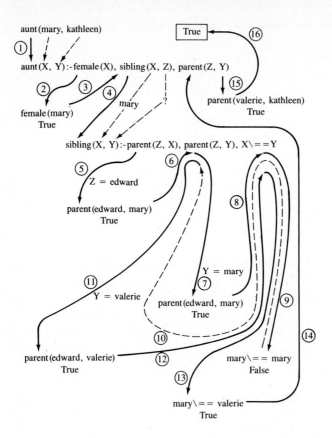

Figure 10-2 Prolog's attempt to prove aunt(mary,kathleen).

sibling(mary,Z), parent(Z,kathleen).

A single head can be unified with the first literal sibling(mary,Z). The result is that the variable X in statement sibling(X,Y) :- parent(Z,X), parent(Z,Y), (X\==Y). is associated with mary, and the variable Y is associated with the variable Z in the goal.

Trying to satisfy the first literal in the latest goal clause

parent(Z,mary), parent(Z,Y), (X\==Y), parent(Y,kathleen)

the system finds the fact parent(edward,mary). Hence edward is **unified** with Z. The goal that the system now attempts to satisfy is parent(edward,Y). The first statement whose head matches the goal is parent(edward,mary). Hence mary is unified with Y. The system now attempts to verify the first literal in the newest goal clause,

(mary\==mary), parent(mary,kathleen).

This literal obviously is not true, and the system backtracks, attempting to find other instantiations of variables that would make a goal *true*. By backing up to the last point where there was more than one match for the first goal, another way to satisfy parent(edward,Y) is tried. This time the statement parent(edward,valerie) is discovered, and valerie is unified with Y. Now the goal clause is

$$(mary\backslash{==}valerie), \text{ parent(valerie,kathleen)}.$$

The first goal is *true* and does not cause new literals to be added to the goal clause. The system tries to verify parent(valerie,kathleen), the only goal left in the goal clause. A fact in the program matches this goal, causing the goal to be removed without adding any new goals. Since the goal clause has been reduced to the empty clause, the original goal—aunt(mary,kathleen)—is *true*, and the system responds yes.

The execution of this program can rightly be viewed as a demand-driven computation. Initially given the goal aunt(mary,kathleen)., the system demands that three subgoals be solved. These goals cause other clauses to be invoked until the original goal is found to be either satisfiable or not satisfiable.

The uses of Prolog extend beyond building and querying data bases; Prolog is a general-purpose programming language. To illustrate this, we present Shapiro's implementation of the insertion sort algorithm [Shapiro 1983b].

```
LINEAR INSERTION SORT (PROLOG):
    sort([X|Xs],Ys) :- sort(Xs,Zs), insert(X,Zs,Ys).
    sort([ ],[ ]).
    insert(X,[Y|Ys],[X,Y|Ys]) :- X <= Y.
    insert(X,[Y|Ys],[Y|Zs]) :- X > Y, insert(X,Ys,Zs).
    insert(X,[ ],[X]).
```

The first argument to sort is the list of atoms to be sorted; the second argument is output of the computation, the sorted list. The first statement says that in order to sort a list of atoms, remove the first element from the list, sort the remainder of the list, then insert the first element into the sorted list. The second statement says that the result of sorting an empty list is the empty list. According to the third statement, an element being inserted into a sorted list goes at the head of the list if its value is less than or equal to the value of the atom at the head of the sorted list. However, if the element X to be inserted into a sorted list [Y|Ys] has a value greater than the value of the first element on the list Y, then according to the fourth statement the inserted list consists of Y followed by the result of inserting X into the remainder of the list Ys. Finally, the fifth statement indicates that the result of inserting an atom into the empty list is a list containing that atom.

Figure 10-3 illustrates the execution of the query sort([3,1,2],X). The final value of X is the output of the computation, the sorted list.

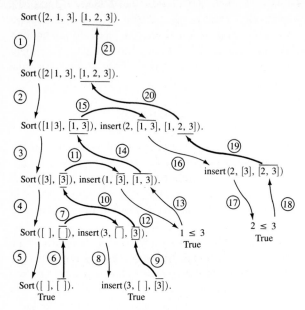

Figure 10-3 Prolog's execution of the query sort([3,1,2],X).

Potential Parallelism in Prolog

The tree corresponding to the search for a solution to a question seems open to various kinds of parallelism. The most obvious technique, called **OR parallelism**, allows processes to search disjunctive subtrees in parallel, reporting back to the parent node the result(s) of the search. In the AND/OR tree of Figure 10-1, for example, the two rules defining aunt could be applied in parallel. The left process would report back the value *true*; the right process would report *false*. Hence the result of the computation is *true*. The advantage of OR parallelism is that the searches are completely independent of each other and may execute concurrently (except that both may share access to a common data base storing facts and rules). The process performing the search of one subtree does not communicate with processes searching other subtrees.

OR parallelism has several disadvantages. First, there can be a combinatorial explosion of goals to be reduced. The system may find itself overwhelmed with work. Second, efficiency is lost when unnecessary goals are reduced, particularly when only a single answer is desired. In the program example above, the leftmost branch has the value *true*, and sequential search would find a solution as fast as the parallel search. A third obstacle to the speedup achievable through OR parallelism is contention among the processes for access to the shared data base of facts and rules.

A second technique, called **AND parallelism**, searches conjunctive subtrees in parallel. In Figure 10-1, for example, the three rules defining sibling(Y.X) could be searched simultaneously. Note, however, that two of these rules share a common variable, Z. When the search reaches each leaf, the data base of facts about relationships is accessed to come up with candidates for Z. The two sets of candidates for Z must be reconciled to see whether any object is in both sets. Hence AND parallelism requires communication and coordination among processes. As in the case of OR parallelism, performing the search in parallel may not improve the running time of the program. (Why?)

A third source of parallelism in Prolog, called **stream parallelism**, is a pipelined form of AND parallelism. The first process finds all unifications for the first term in the body of the rule. The results of each unification are sent to the process handling the second term in the rule body, and so on. Clearly the parallelism achievable is dependent upon the number of unifications possible and the number of terms in the rule body. As with all pipelined algorithms, there is pipe filling and emptying overhead.

The three sources of parallelism in the execution of logic programs described thus far are parallelizations of the search process. A fourth source is **parallel unification**, the parallel identification of all possible unifications for a particular goal. This approach has not been discussed as widely in the literature.

10-3 CONCURRENT PROLOG

As we saw in the previous section, Prolog performs a sequential depth-first search for a proof. The order in which goals are considered is determined by the order of clauses in a program and the order of goals within a clause. Concurrent Prolog is an extension of Prolog that allows parallelism in the search for a proof. Concurrent Prolog is of special interest because it is the basis of the Kernel Language of the Parallel Inference Engine being developed at the Institute for New Generation Computer Technology (ICOT) in Japan.

A unit goal corresponds to a process, while a conjunctive goal corresponds to a system of related processes. Processes are created when goals are reduced; processes die when they are reduced to *true*, the empty goal.

The parallel processes are controlled through the use of two additions to Prolog: guarded-command indeterminacy and data flow synchronization. Guarded-command indeterminacy allows a process to explore some of the goals in the body of a clause in parallel before deciding whether to pursue the remaining goals. Data flow synchronization is implemented through the use of a "read only" annotation on variables. Occurrences of variables can be labeled as "read only"; a process attempting to instantiate such a variable suspends execution until that variable is instantiated by another process.

A Concurrent Prolog program consists of a finite number of guarded clauses of the form

$$H \text{ :- } G_1, \ldots, G_m \mid B_1, \ldots, B_n. \quad m, n \geq 0.$$

H is the head of the clause, $G_1, \ldots, G_m$ is the guard of the clause, and $B_1, \ldots, B_n$ is the body of the clause. The "|" is the guard symbol. If the guard is empty (that is, $m = 0$), the guard symbol is omitted.

Declaratively, a guarded clause can be interpreted "H is *true* if the Gs and the Bs are *true*." Procedurally, a guarded clause can be interpreted "To reduce goal A, unify A with H and, if successful, reduce the Gs to the empty system and, if successful, commit to that clause and, if successful, reduce A to the Bs."

In the execution of a Concurrent Prolog program, when a goal A needs to be satisfied, the guards of all the guarded clauses whose heads unify with A are executed concurrently. Within a single guard there may be a number of conjunctive goals. These goals execute in parallel in the same environment. There is no communication between the processes associated with different guarded clauses, however; they operate in independent environments.

When a guard succeeds, an attempt is made to unify the guard's environment with A. If this commitment succeeds, then A is reduced to the body of the guarded clause. The commit operator has two purposes. First, it ensures that A is reduced to the body of only one guarded clause. As soon as a single guard attempts to commit, all the processes associated with other guarded clauses whose heads unified with A are terminated. Hence Concurrent Prolog yields a single solution to any goal. The second purpose of the commit operator is to export the bindings of the variables in the guard to other processes sharing those variables.

To summarize, Concurrent Prolog makes use of both OR parallelism (simultaneous execution of every guarded clause whose head unifies with the goal) and AND parallelism (simultaneous execution of goals within a guard).

Synchronization of processes and message passing may be accomplished by the read-only annotation mentioned earlier. If a process attempts to unify a variable marked with the read-only annotation "?," then that process suspends execution until another process instantiates that variable, i.e., gives it a value. Logical variables can be either uninstantiated or instantiated. Once it is instantiated, the value of a logical variable cannot be modified. Hence a logical variable can act as a communication channel that transmits a single message.

A wide variety of algorithms have been implemented in Concurrent Prolog, including Shiloach and Vishkin's maximum network flow algorithm, a parallel parsing algorithm, an OR parallel Prolog interpreter, an object-oriented knowledge representation language, and a variety of systolic "number crunching" algorithms.

10-4 THE BAGEL

The Bagel is a parallel architecture developed by Shapiro and his associates at the Weizmann Institute of Science in Israel [Shapiro 1983c]. The purpose of this section is twofold: to review questions that should be raised by the designer of any new parallel computer and to present Shapiro's response to these questions, as embodied in the Bagel. You should remember that there are no generally accepted answers to these fundamental questions.

Should the computer be massively parallel? We have seen some strong arguments against the utility of computers containing a massive number of CPUs. Some would say that such an architecture cannot be general purpose. Shapiro does not agree. Assuming Shapiro is right, what are the consequences of choosing to implement a massively parallel architecture? First, the architecture must be **scalable**; that is, increasing the number of processors must produce a corresponding increase in the processor power of the machine, and there should be no upper limit to the number of processors used.

Clearly a scalable architecture cannot have global memory, because you cannot put an unbounded number of processors close to a global memory, and if processors are put farther and farther from the memory, then performance will suffer. If there is no global memory, then communication costs must be carefully considered. We have already seen that many algorithms are dominated by their communication costs. When a multicomputer is programmed, care must be taken to ensure that virtually all communication is between processors that are close together. For this reason systolic algorithms have a great deal of appeal, because they do a good job of balancing computation and local communication costs.

Should the user of the parallel computer be aware of the underlying architecture? To execute efficiently, programs must control their use of space, time, and communication. That means there must be some mechanism to map processes to processors. Currently there does not exist any automated means for allocating processes to processors, and the program designer must do the mapping. Thus the user must be aware of the computer's architecture.

Whether it is the programmer or the compiler that is mapping processes to processors, there must be some sort of notation to express it. To make the process-to-processor mapping problem tractable, the processor interconnection scheme must be simple and regular.

The Bagel is a systolic Concurrent Prolog architecture that reflects the above assumptions. It is a multicomputer with a two-dimensional mesh-connected topology and shifted end-around toroidal interconnections (Figure 10-4). This interconnection topology has several advantages. First, it allows the programmer to assume that there is a virtual infinite two-dimensional processor space. Second, it supports an even mapping of processes to processors, since any path in a single direction visits every processor exactly once before returning to the original processor. Third, it is easy to construct. Fourth, it is scalable.

The kernel language of the Bagel is Concurrent Prolog enhanced with a mapping notation called **Turtle programs**. Like the Turtle in the language LOGO, each Concurrent Prolog process has a position and a heading. Turtle programs are used to specify which processor a process is to run on. Recall that goals correspond to processes in Concurrent Prolog. On this system goals can take the form

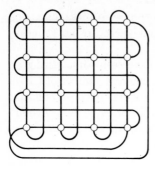

Figure 10-4 A two-dimensional mesh with shifted end-around toroidal interconnections.

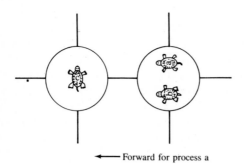

← Forward for process a

Figure 10-5 The result of executing the enhanced Concurrent Prolog clause a(X,Y) :- b(X),c(Y)@forward,right.

Goal @ TP

where TP is a Turtle program.

A Turtle program is a sequence of one or more of the following instructions (among others): forward(⟨distance⟩), back(⟨distance⟩), left(⟨angle⟩), and right(⟨angle⟩). The terms enclosed in angle brackets are arguments to be specified by the programmer. Default values are 1 for ⟨distance⟩ and 90 for ⟨angle⟩. For example, the enhanced Concurrent Prolog clause

a(X,Y) :- b(X),c(Y)@forward,right.

would, when executed, spawn two new processes. The process corresponding to the goal b(X) would have the same position and orientation as the process for goal a(X,Y). The process for goal c(Y) would have position one processor ahead of the processor containing processes A(X,Y) and b(X), and its orientation would be a quarter turn clockwise of the orientation of the other two processes (Figure 10-5).

As we have already seen, systolic algorithms seem to be a good match for this kind of architecture. Shapiro has implemented a good number of systolic algorithms on a simulator for the Bagel. What follows is a systolic insertion sort algorithm. This algorithm is similar to the insertion sort algorithm described previously, but it has been written in Concurrent Prolog enhanced with Turtle graphics, and hence parallelism in the algorithm can be exploited.

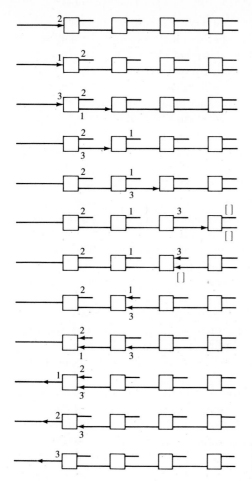

Figure 10-6 Execution of systolic insertion sort program.

PARALLEL INSERTION SORT (BAGEL):
```
sort([X|Xs],Ys) :- insert(X,Zs?,Ys),sort(Xs?,Zs)@forward.
sort([ ],[ ]).
insert(X,[Y|Ys],[Y|Zs?]) :- X>Y | insert(X,Ys?,Zs).
insert(X,[Y|Ys],[X,Y|Ys]) :- X=<Y | true.
insert(X,[ ],[X]).
```

The execution of this algorithm is illustrated in Figure 10-6. Parallelism is possible in this algorithm because lists are streamed from process to process. Hence this algorithm can sort $O(n)$ names in $O(n)$ time using n processors.

10-5 SUMMARY

The purpose of the Japanese fifth-generation computing project is to develop high-speed knowledge-processing parallel computers capable of performing billions of logical inferences per second. A number of trends in computer science suggest that logic programming may be the best method to implement such a system, although many do not share this viewpoint. Prolog has proved to be a successful logic programming language; Concurrent Prolog is an enhancement of Prolog allowing searches to be performed in parallel. Concurrent Prolog is the basis for the kernel language of ICOT's parallel inference engine; it is also the machine language of the Bagel, an experimental parallel computer that combines the ideas of massive parallelism, logic programming, and systolic algorithms.

BIBLIOGRAPHIC NOTES

The preliminary report on fifth-generation computer systems appears in Motooka [1982]. Journals concentrating on this topic include *Future Generations Computer Systems* and *New Generation Computing*. Kowalski [1979] and Lloyd [1984] have written books on the subject of logic programming. An early description of Prolog was written by Warren, Pereira, and Pereira [1977]. A standard reference for the Prolog programming language is Clocksin and Mellish [1981]. I am indebted to Ehud Shapiro and the Weizmann Institute of Science for providing me with valuable technical reports on Concurrent Prolog and the Bagel [Mierowsky et al. 1985; Shapiro 1983c, 1985].

Conery and Kibler [1984] discuss AND parallelism in logic programs. Lindstrom and Panangaden [1984] write about stream parallelism. Douglass [1984] has written a very readable paper critiquing approaches to the parallelization of rule-based expert systems, including production systems as well as systems based on logic programming. His paper highlights the lack of empirical evidence that could help determine the parallelizability of expert systems.

EXERCISES

10-1 Define the term *logical inferences per second*.

10-2 In the programming language Prolog, what is the difference between a fact and a rule?

10-3 Give an example in which the use of AND parallelism does not improve the time needed to search for a solution and may, in fact, increase the search time.

10-4 Show how the computation grandparent(X,bill). would be performed, given the following Prolog program:

```
father(don,eric).
mother(jane,eric).
father(eric,bill).
mother(ruth,bill).
```

```
grandparent(X,Y) :- parent(X,Z),parent(Z,Y).
parent(X,Y) :- mother(X,Y).
parent(X,Y) :- father(X,Y).
```

How many logical inferences are performed in this computation?

10-5 Draw a picture of a 36-processor Bagel, keeping all interprocessor connections roughly the same length.

10-6 Is the Bagel a data flow computer? Justify your answer.

ELEVEN

PIPELINED VECTOR PROCESSORS

In Chapter 1 we saw that there has been a natural evolution of computer architectures from the highly serial first-generation computers to the highly pipelined Cray-1. Although the emphasis of this book has been on multiprocessors, multicomputers, and processor arrays, there are two good reasons why pipelined vector processors are worth some study. First, most supercomputers in existence are pipelined vector computers. Second, some multiprocessors, such as the Cray-3, are actually multiple pipelined vector processors. Hence it is important for us to understand (albeit at a high level) how pipelined vector processors work.

This chapter begins by outlining the vector-processing capabilities that distinguish vector computers from serial computers. Pipelined vector processors are one way to implement vector computers. We discuss the difference between register-to-register and memory-to-memory architectures. Sections 11-3 and 11-4 describe two pipelined vector processors: the Cray-1 and the Cyber-205. How can the performance of these computers be measured and compared? One way is to use the metrics developed by Hockney and Jesshope. Section 11-5 discusses the performance of these computers on several algorithms in terms of Hockney and Jesshope's notation. Section 11-6 is devoted to vectorizing compilers, compilers that are able to convert blocks of sequential instructions into vector instructions that can be pipelined. The existence of vectorizing compilers has made the programmer's task easier, but good algorithm design still has an important role. Techniques for designing algorithms for pipelined vector processors are discussed in Section 11-7. This chapter concludes with a description of attached processors, low-cost alternatives to pipelined vector computers.

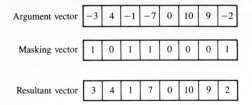

Figure 11-1 Use of a masking vector to compute $V \leftarrow |V|$ by negating selected elements.

11-1 VECTOR COMPUTERS

The instruction set of a serial computer allows the manipulation of scalar operands. A **vector computer** is a computer with an instruction set that includes operations on vectors as well as scalars.

A vector is an ordered set of homogeneous elements. Examples of vectors include one-dimensional arrays and individual rows, columns, and diagonals from two-dimensional arrays. Vectors are made up of scalar quantities such as integers, floating-point numbers, characters, or booleans. Vector instructions generally fall into one of three categories:

$$V \leftarrow V$$
$$V \leftarrow V \odot V$$
$$V \leftarrow S \odot V$$

where V and S represent vector and scalar operands and $\odot$ represents an operation. An example of a $V \leftarrow V$ instruction is vector square root. Each element of the resultant vector is the square root of the corresponding element of the original vector. Vector addition is an example of an instruction of the $V \leftarrow V \odot V$ category. Vector-scalar multiplication fits into the $V \leftarrow S \odot V$ category; each element of the resultant vector is assigned the product of a scalar and the corresponding element of the input vector.

Some authors call operations of the form $S \leftarrow V$ vector operations. Vector summation is an example of such an operation. We do not agree with this categorization for two reasons. First, the operation has a scalar result. Second, it cannot be pipelined.

Often an operation needs to be performed on particular elements of a vector. Two vectors can be compared (or a vector compared with a scalar), producing a boolean vector. This boolean vector can then be used as a **masking vector** to enable or disable operations on particular vector elements (Figure 11-1).

What if an operation is to be performed on a very small percentage of the elements of a vector? A masking vector could be used, as described earlier, but it is wasteful to manipulate a long vector when so little work is to be done. An alternative is to use a **compress** instruction, which loads a vector according to the values in a corresponding masking vector. At this point the compressed vector can be manipulated as an operand in vector instructions. The inverse operation of compress, called **expand**, is used to store a vector according to the

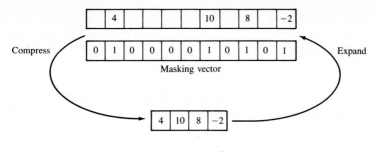

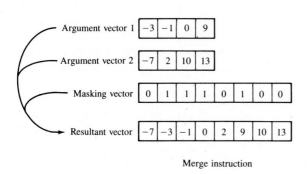

Figure 11-2 Compress, expand, and merge instructions.

values in a corresponding masking vector. Another special instruction, **merge**, merges two vectors according to the values in a masking vector. Figure 11-2 illustrates the compress, expand, and merge instructions.

Two other vector manipulations can pose problems for vector computers, because they do not access memory locations in a uniform manner. The operations, called **scatter** and **gather**, are most easily explained by their implementation in a high-level language:

Scatter:

```
for  i  ← 1 to n do
    a[index[i]]  ←  b[i]
endfor
```

Gather:

$$\text{for } i \leftarrow 1 \text{ to } n \text{ do}$$
$$a[i] \leftarrow b[index[i]]$$
$$\text{endfor}$$

Some pipelined vector processors are able to perform some of these operations in hardware; the rest must perform them in software. For example, the Fujitsu VP-200 has compress, expand, scatter, and gather implemented in hardware. Scatter and gather instructions are efficiently implemented by microcode in the stream unit of the Cyber-205.

11-2 PIPELINED VECTOR PROCESSORS

It is natural to implement vector computers as pipelined processors. In vector instructions the same operation is performed on a large number of elements. Consider the high-level language program segment in Figure 11-3 and its "machine-language" equivalent. Look at the number of instructions devoted to loop control. Pipelining the data stream can eliminate this loop control overhead. Pipelines do have a startup delay, but as vectors become long, this overhead becomes less and less significant. It is common for applications to manipulate vectors with 50,000 or more elements.

A pipelined vector processor can be classified as a memory-to-memory architecture or a register-to-register architecture, depending upon how vector instructions are executed. In a memory-to-memory architecture, vectors are streamed directly from memory to the arithmetic units and back to memory. In a register-to-register architecture internal registers are used to store operands and results. The Cyber-205 has a memory-to-memory architecture, while the Cray-1 has a register-to-register architecture.

```
              DO 10 I = 1, N
        10      A(I) = B(I) + C(I)

    STORE         I ← 1
    GOTO          Z
 Y: LOAD          B[I]
    LOAD          C[I]
    ADD           B[I] + C[I]
    STORE         A[I] ← B[I] + C[I]
    INCREMENT     I ← I + 1

 Z: IF I ≤ N GOTO Y
```

Figure 11-3 FORTRAN DO loop and machine-language equivalent.

Pipelined vector processors have three overheads to contend with in performing vector operations. First, it takes time to fill the pipe. Until the pipe is full, the computer is not working at full speed. Second, it takes time to drain the pipe. Third, if the computer is based on a register-to-register architecture and a vector in memory is longer than a vector register, then there is an overhead involved in breaking a vector into pieces small enough to fit into the vector registers.

Memory Organization

Recall that memory interleaving is the distribution of memory addresses across a number of memory banks (modules). In high-order interleaving each memory bank contains a block of consecutive addresses, while in low-order interleaving consecutive addresses are in different memory banks. A memory conflict occurs when a processor tries to access a location in a memory bank before a previous access to that bank has been completed. The time needed to access a memory location is called the **memory cycle time**.

High-order interleaving would be disastrous on a pipelined vector processor, because it would lead to excessive memory conflicts. Since the memory cycle time is much greater than the processor clock cycle time, the placement of a vector in a single memory bank would cripple attempts at pipelining. Hence pipelined vector processors use low-order interleaving to reduce memory conflicts.

11-3 THE CRAY-1

The Cray-1 is the most popular pipelined vector processor ever built. The first Cray-1 was delivered to Los Alamos National Laboratory in 1976; over 60 have now been shipped.

A functional diagram of the Cray-1 is shown in Figure 11-4. Twelve independent pipelined functional units perform the arithmetic and logic operations. Since the Cray-1 is a register-to-register architecture, the functional units take data from and put results into registers.

Memory is divided into 16 memory banks that may operate concurrently. Memory has a 50-nanosecond access and cycle time. Since the clock period of the processor is 12.5 nanoseconds, it takes four clock cycles for a memory bank to handle a request. There is a 64-bit-wide data bus between main memory and the registers. If each word is drawn from a different memory bank, this bus can transfer one 64-bit word from memory to a register per clock cycle, yielding a memory bandwidth of 80 megawords per second. If a sequence of words is drawn from the same memory bank, then the bus can transfer one 64-bit word from memory every four clock cycles, resulting in a memory bandwidth of 20 megawords per second.

A typical arithmetic operation requires three memory accesses. For example, three memory accesses (two loads and a store) are required to add two arguments

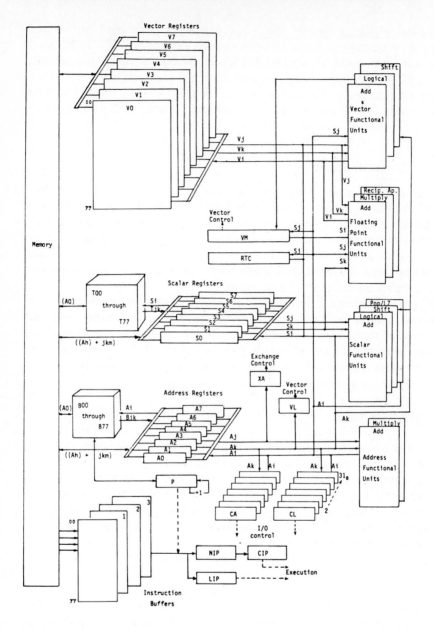

Figure 11-4 Architecture of the Cray-1. (Reprinted by permission of Cray Research.)

stored in memory and store the result. The maximum computing rate of the Cray-1 is 160 megaflops (80 million additions and 80 million multiplications per second). To support this high rate of computation, the memory bandwidth would have to be 480 megawords per second. As we have seen, the Cray-1 has a maximum memory bandwidth (transferring data) of 80 megawords per second. This low register-to-memory bandwidth can be a bottleneck to high performance on the Cray-1. One partial remedy for this problem is to carefully code programs so that most of the computations are between data already in registers.

Strip mining is the process of dividing a DO loop into a number of smaller loops that are executed sequentially, due to limitations in the length of vector registers in a register-to-register pipelined vector processor. For example, suppose a single FORTRAN DO loop performs a pairwise addition on two vectors of length 100. Since the vector registers in the Cray-1 can hold at most 64 elements, the DO loop must be performed in two steps. Step 1 consists of loading the first 64 elements from each vector, performing the first 64 additions, then storing the first 64 results. Step 2 consists of loading the final 36 elements from each vector, performing the final 36 additions, and storing the final 36 results.

The data bus cannot transfer information from memory to a register and from a register to memory at the same time. The flow of data must be in the same direction. The length of this data bus "pipeline" between memory and the registers is 11 clock periods.

The Cray-1 has four instruction buffers. Each instruction buffer holds up to 64 instructions of 16 bits, and each instruction buffer is connected to memory by a 64-bit-wide bus. Hence the data bus between main memory and an instruction buffer can transfer a 64-bit-word every 50 nanoseconds, and the total memory bandwidth is 320 megawords per second, assuming the instruction buffers are retrieving instructions from different memory banks.

The CPU contains eight 24-bit address registers, eight 64-bit floating-point scalar registers, and eight floating-point vector registers. Each vector register can hold up to 64 floating-point numbers of 64 bits. There are sixty-four 24-bit buffer registers between the address registers and main memory and sixty-four 64-bit registers between the scalar registers and main memory.

The 12 pipelined functional units work with data in the address, scalar, vector, and vector mask registers. Each functional unit may accept a new set of arguments every clock period. The vector functional units operate on arguments from a pair of vector registers or from a vector register and a scalar register. The result can be put in either a vector register or a vector mask register. Additionally, the Cray-1 can chain together a series of vector operations, allowing a second vector operation to begin as soon as results begin streaming out of the first functional unit's pipeline. The advantage of chaining is illustrated in Figure 11-5, which shows the marked reduction in execution time that can be achieved when a second vector computation can begin before the first has completed.

The Cray X-MP

The Cray X-MP, released in 1983, is an improved version of the Cray-1. Specifically, the Cray X-MP has a clock cycle of 9.5 nanoseconds, improved memory bandwidth, an increase in the maximum memory size, and the capability of having one, two, or four pipelined vector processors. Since the pipelined processors

DO 10 I = 1, N
10   Z(I) = (W(I) + X(I)) * Y(I)

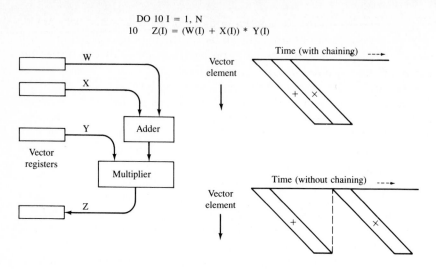

Figure 11-5 Chaining vector operations on the Cray-1.

are able to cooperate on a single computation, the X-MP is a tightly coupled multiprocessor.

11-4 THE CYBER-205

Control Data Corporation manufactures the Cyber-205. Its first delivery of the computer was to the United Kingdom Meteorological Office in 1981.

The Cyber-205 has either two or four memory units, each containing 1 million 64-bit words. Each 1-megaword memory unit has a 512-bit-wide data path connecting it with the memory interface. Figure 11-6 is a diagram of a memory unit on the Cyber-205. A 512-bit entity is called a **sword** (for superword). Each data path can transfer data at the rate of 1 sword every clock period (20 nanoseconds). Hence memory bandwidth is 400 megawords per second per million words of memory. Each 1 megaword of memory is divided into 16 memory stacks; each of these 16 stacks has a 32-bit stack data path to the memory interface ($16 \times 32 = 512$). Consecutive addresses are stored in different stacks; hence one sword can be accessed in parallel. Each stack is divided into eight memory banks. Memory cycle time is four clock periods, or 80 nanoseconds.

The architecture of the Cyber-205 is shown in Figure 11-7. The scalar portion of the CPU consists of a load/store unit and five arithmetic functional units. The functional units take arguments and return results to a set of 256 registers of 64 bits. All the functional units are pipelined and may take new arguments every clock period, with the exception of the unit that does division, square root, and conversions. Concurrency is limited by the fact that the registers can supply at most one pair of arguments to the functional units and accept one result every

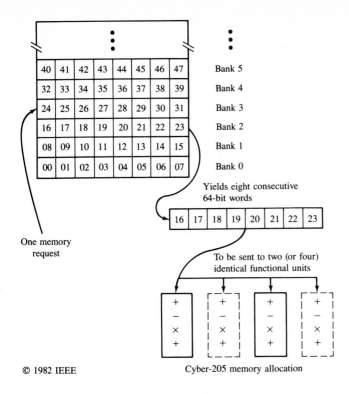

Cyber-205 memory allocation

Figure 11-6 Memory unit on the Cyber-205. (Lincoln [1982].)

clock period. However, the result of any functional unit can be passed directly to another functional unit without first being sent to a register. Using a result in this way is called **shortstopping**. The load/store unit moves data between these registers and memory.

A **stream unit** feeds vectors into the vector arithmetic section of the CPU, which consists of one, two, or four floating-point pipelines and a string unit. Since the Cyber-205 has no vector registers, argument vectors are fetched from memory, and the resultant vectors are stored in memory. This means that the overhead for vector operations is larger on the Cyber-205 than on the Cray-1. Vectors may contain up to 65,535 elements.

The string unit is used to process boolean mask vectors. This allows operations to be performed on selected elements of vectors. In addition, scatter and gather operations are microcoded in the stream unit.

Each floating-point pipeline consists of five separate pipelined functional units: an addition unit, multiplication unit, shift unit, logical unit, and delay unit. The multiplication unit is used for division. Each unit can generate results at the rate of 50 million results per second for 64-bit operations and 100 million results per second for 32-bit operations. If two successive vector instructions use different units and contain one scalar operand, then linkage can take place. The

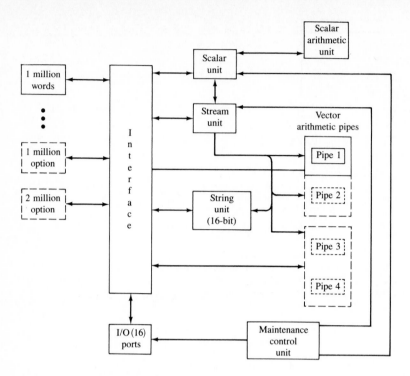

Figure 11-7 Architecture of the Cyber-205. (Courtesy of Control Data Corporation.)

output of the first functional unit can be made the input of the second functional unit, avoiding a trip to main memory and back. The function of these **linked triadic operations** is the same as chaining on the Cray-1, but their use is more restricted, since only two operations can be linked and one of the operators must be a scalar. Examples of linked triadic operations include

$$V \leftarrow V + S \times V \qquad V \leftarrow (V + S) \times V$$

The use of linked triadic operations increases the maximum performance of a pipelined functional unit to 100 million results per second for 64-bit operations and 200 million results per second for 32-bit operations. Thus the maximum asymptotic performance of a Cyber-205 with four pipelined functional units is 400 megaflops in 64-bit arithmetic and 800 megaflops in 32-bit arithmetic.

11-5 PERFORMANCE MEASUREMENT

Hockney and Jesshope [1981] have proposed the use of two parameters to provide a rough description of the performance of computers. Although they intend these parameters to be definable for any real computer, the parameters are most easily used in the context of serial and vector computers.

The two parameters are referred to as r_∞ and $n_{1/2}$. The parameter r_∞ is defined to be the maximum computation rate of the computer expressed in units of equivalent scalar operations performed per second. This parameter is primarily a function of the existing computer technology: as computers get faster, r_∞ increases.

The second parameter, $n_{1/2}$, is the vector length at which the computer reaches half its maximum performance. This parameter indicates the amount of parallelism in an architecture. For example, a serial computer would have $n_{1/2} = 0$, while an infinite array of processors would have $n_{1/2} = \infty$.

The length of the pipeline depends upon the operation being carried out. Hence the value of $n_{1/2}$ is not a constant, even for a given computer. Hockney and Jesshope also point out that any unnecessary overhead introduced into loop control by a compiler will increase the value of $n_{1/2}$. Nevertheless, $n_{1/2}$ gives at least some idea of the vector performance of a computer.

The half-performance length and maximum performance of a computer are determined experimentally. If the time needed to execute a program segment on various sized vectors is plotted against the vector length n, then the intercept of the line with the n axis gives the value $-n_{1/2}$, and the slope of the line gives the value of $1/r_\infty$.

Consider, for example, the graph in Figure 11-8, measuring the time needed to perform element-by-element multiplication of two vectors of length n on a Cray-1 using vectorized FORTRAN. (The timings were done by Craigie and reported in Hockney and Jesshope [1981].) The n axis represents the number of floating-point operations performed; the y axis represents the time in microseconds. The solid line represents the time measured by Craigie. Note the jumps in the curve at $n = 65$, $n = 129$, and $n = 193$. These jumps represent the time needed to reload the vector registers. The dashed line, representing the average

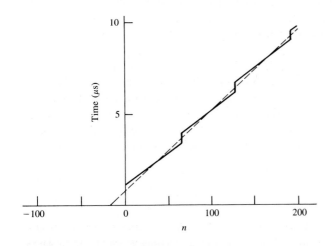

Figure 11-8 Measurement of r_∞ and $n_{1/2}$ on the Cray-1 performing elementwise vector multiplication. (Hockney and Jesshope [1981].)

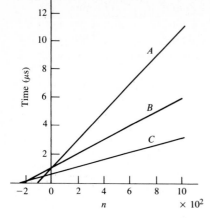

Figure 11-9 Measurement of r_∞ and $n_{1/2}$ on the Cyber-205. Curve A—two pipes, vector addition, 64-bit values. Curve B—two pipes, vector addition, 32-bit values. Curve C—four pipes, triadic operation, 64-bit values. (Hockney and Jesshope [1981].)

slope, is used to determine r_∞ and $n_{1/2}$. The slope of the dashed line is 4.55 microseconds per 100 flops. Hence $r_\infty = 100$ flops per 4.55 microseconds = 22 megaflops. The n intercept is -18; hence $n_{1/2} = 18$.

Now examine the graph in Figure 11-9. Curve A represents the time needed to perform dyadic vector addition on a two-pipe Cyber-205 doing 64-bit arithmetic: $r_\infty = 100$ megaflops and $n_{1/2} = 100$. Curve B is a timing curve for a two-pipe Cyber-205 performing dyadic vector addition with 32 bits of accuracy or a four-pipe Cyber-205 performing dyadic vector addition on 64-bit numbers: $r_\infty = 200$ megaflops and $n_{1/2} = 200$. Finally, curve C shows the time needed for a four-pipe Cyber-205 to perform a linked triadic vector operation on 64-bit numbers: $r_\infty = 400$ megaflops and $n_{1/2} \approx 200$.

Figure 11-10 contrasts the performance of the two pipelined vector processors we have discussed executing a FORTRAN statement of the form

```
V1(I) = S1 * V2(I) + S2 * V3(I)
```

This particular statement was chosen because it seems to require a typical number of loads and stores per floating-point operation [Bucher and Simmons 1985]. However, do not assume that this single benchmark is a complete, or even fair, characterization of the relative performance of these computers. The primary purpose of Figure 11-10 is to reinforce the concepts of r_∞ and $n_{1/2}$. The maximum speed of the computer, r_∞, is inversely proportional to the slope of the execution time curve. The half-performance vector length of the machine is directly proportional to the intercept on the n (horizontal) axis. On this particular computation, which computer has the highest value of r_∞? Which computer has the smallest value of $n_{1/2}$?

Vector operations fall into one of three categories. Some vector operations manipulate vectors stored in contiguous memory locations. A second category of vector operation manipulates vector elements stored a constant distance (stride) apart, for example, every seventh memory location. The third type of vector

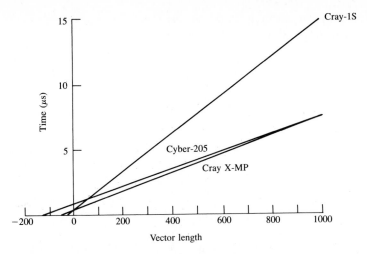

Figure 11-10 Contrasting the performance of the Cray-1, Cray X-MP, and Cyber-205 executing a FORTRAN statement of the form V1(I) = S1 * V2(I) + S2 * V3(I).

operation manipulates vector elements randomly distributed through memory. Table 11-1 presents typical vector speeds for each category of vector operation, for the Cray-1S, the single-processor Cray X-MP, the two pipeline Cyber-205, and the four pipeline Cyber-205. Assuming that 78 percent of vector operations involve contiguous vectors, 20 percent of vector operations involve constant strides, and 2 percent of vector operations require random memory access, the table also presents the effective vector speed of these computers. (The performance measurements quoted in the previous figure and table appear in Bucher and Simmons [1985].)

11-6 VECTORIZING COMPILERS

It is preferable to have vector constructs in high-level languages; then the programmer and the compiler are on the same side. Unfortunately, most programs written for pipelined vector processors are written in sequential languages, such as FORTRAN and Pascal. The job of a **vectorizing compiler**, then, is to compile a program written in a sequential language while converting as many blocks of sequential operations as possible into vector instructions that can be pipelined.

It is easy to see how the statements

```
DO 10 I = 1, N
    X(I) = Y(I) + Z(I)
10 CONTINUE
```

can be translated into a vector instruction in object code. Other times, however,

Table 11-1 Vector speeds in megaflops for pipelined vector processors, assuming vector length 100 and 64-bit arithmetic

Computer	Vector speed for contiguous vectors	Vector speed for constant stride	Vector speed for random memory access	Effective vector speed
CDC Cyber-205 (two vector pipes)	64	19	12	41
CDC Cyber-205 (four vector pipes)	87	19	12	47
CRI Cray-1S	58	56	5	46
CRI Cray X-MP/24 (one of two processors)	116	97	6	83
CRI Cray X-MP/48 (one of four processors)	116	97	32	106

the translation is not as straightforward. The interesting vectorizations that follow can be found in Hwang and Briggs [1984].

This DO loop contains no vector instructions:

```
   DO 10 I = 1, N
      X(I) = Y(I-1)
      Y(I) = 2*Y(I)
10 CONTINUE
```

However, this set of two DO loops is functionally equivalent and clearly vectorizable:

```
   DO 10 I = 1, N
      Y(I) = 2*Y(I)
10 CONTINUE
   DO 20 I = 1, N
      X(I) = Y(I-1)
20 CONTINUE
```

Sometimes an otherwise unvectorizable section of code can be vectorized by using a temporary vector. For example, the way to vectorize the statements

```
   DO 10 I = 1, N
      X(I) = Y(I) + Z(I)
      Y(I) = 2*X(I+5)
10 CONTINUE
```

is clear once they are rewritten as

```
    DO 10 I = 1, N
       TEMP(I) = X(I+5)
   10 CONTINUE
    DO 20 I = 1, N
       X(I) = Y(I) + Z(I)
   20 CONTINUE
    DO 30 I = 1, N
       Y(I) = 2*TEMP(I)
   30 CONTINUE
```

Sometimes more than one opportunity arises for vectorization, and the compiler must be able to determine which option has the potential for the highest performance. For example, it is better for the length of the vector to be as long as possible. Hence the FORTRAN statements

```
    DO 10 I = 1, 100
       DO 20 J = 1, 10
          X(I,J) = Y(I,J) * Z(I,J)
   20     CONTINUE
   10 CONTINUE
```

are better vectorized after being rearranged into

```
    DO 10 J = 1, 10
       DO 20 I = 1, 100
          X(I,J) = Y(I,J) * Z(I,J)
   20     CONTINUE
   10 CONTINUE
```

11-7 ALGORITHM DESIGN

Vectorizing compilers are becoming quite sophisticated. However, it is not yet (and may never be) true that the programmer can ignore the architecture completely, code a sequential algorithm in any form whatsoever, and expect the vectorizing compiler to generate a compiled program that takes full advantage of the power of the computer. For example, since a pipeline is used to process scalar quantities, it is better to perform as many scalar instructions as possible at one time, reducing the overhead due to pipeline reconfiguration. The remainder of this section illustrates the development of two algorithms for the Cray-1.

Matrix Multiplication

The product of an $l \times m$ matrix $\mathbf{A}$ and an $m \times n$ matrix $\mathbf{B}$ is an $l \times n$ matrix $\mathbf{C}$ whose elements are defined by

$$c_{i,j} = \sum_{k=0}^{m-1} a_{i,k} b_{k,j}$$

A sequential algorithm implementing matrix multiplication appears below:

```
MATRIX MULTIPLICATION (INNER PRODUCT/SISD):
begin
    for i ← 0 to l − 1 do
        for j ← 0 to n − 1 do
            c_{i,j} ← 0
            for k ← 0 to m − 1 do
                c_{i,j} ← c_{i,j} + a_{i,k} × b_{k,j}
            endfor
        endfor
    endfor
end
```

The innermost loop computes the inner product of the ith row of $\mathbf{A}$ and the jth column of $\mathbf{B}$; hence this algorithm is called the **inner-product method.**

The inner-product method performs the n^2 inner product calculations sequentially. Consider what happens if the i loop is made the innermost loop. The following algorithm results.

```
MATRIX MULTIPLICATION (MIDDLE PRODUCT/SISD):
begin
    for i ← 0 to l − 1 do
        for j ← 0 to n − 1 do
            c_{i,j} ← 0
        endfor
    endfor
    for j ← 0 to l − 1 do
        for k ← 0 to m − 1 do
            for i ← 0 to n − 1 do
                c_{i,j} ← c_{i,j} + a_{i,k} × b_{k,j}
            endfor
        endfor
    endfor
end
```

This algorithm computes the inner products for an entire column of $\mathbf{A}$ at a time; it is called the **middle-product method.** What is the advantage of the second algorithm over the first? Notice that the addition in the first algorithm is a scalar addition, while the addition in the second algorithm is actually a vector addition; in other words, the innermost **for** loop of the second algorithm is actually a vector operation of the form

$$V \leftarrow V + V \times S$$

Indeed, the middle-product method, when it is programmed in assembly language, is the fastest way to multiply matrices on the Cray-1 [Hockney and Jesshope 1981].

One-Dimensional Table Lookup

Given a value x and a table of increasing values $t_1, t_2, t_3, \ldots, t_{n-1}, t_n$, the one-dimensional table lookup problem is to find the smallest integer $i = i(x)$ such that $x < t_i$ [Dubois 1982]. To borrow Dubois' example, the speed of sound is no simple function of altitude. Hence the speed of sound at various altitudes is often kept in a table, and the speed of sound at intermediate altitudes is estimated through interpolation. A fast table lookup is necessary to determine what table altitudes straddle the altitude in question.

Before reading further, decide how you would solve this problem on a sequential computer. Then try to devise an algorithm suitable for a pipelined vector processor. Note that for realistic problems, table sizes usually range from 5 to 500 elements.

A sequential algorithm with asymptotically optimal complexity is bisection, with complexity $O(\log n)$. Unfortunately, the key steps of this algorithm—fetching the bisecting element from the table, comparing its value with the value in question, and deciding whether to modify the upper or the lower bound—are not vector instructions and cannot take full advantage of the power of a pipelined vector processor. Two other algorithms are of interest.

Rather than comparing a single element of t with x, why not compare a large number of elements? This idea is the basis for the second algorithm, called *linear vector scan*. Assuming a register-to-register architecture with R registers, this algorithm makes R comparisons in a single pass. Hence the first iteration of the algorithm would compare t_1 through t_R with x. If all the elements are smaller than x, then t_{R+1} through t_{2R} would be compared with x in the second iteration, and so on. Eventually during an iteration not all the elements will be smaller than x, and given that a bit mask is the result of every iteration—where a 1 bit represents a value smaller than x and a 0 bit represents a value larger than x—then the number of bits set in the bit mask represents the index of $i(x)$ in the group of R elements. The linear vector scan algorithm has complexity $\Theta(n)$, but Dubois has calculated that on a Cray-1 it is faster than bisection for $n \le 256$, i.e., on tables of reasonable size.

The third algorithm, called *M-section*, combines the virtues of the first two algorithms. Bisection has the advantage that it requires no more than $\log n$ iterations. Linear vector scan is able to take advantage of the speed of the computer performing vector operations. *M-section* compares a number of elements of T with x, but these elements are distributed among T so that only a logarithmic number of iterations are required.

Table 11-2 Average time to perform table lookup on Cray-1

	Time ($\mu s/n$)	
Vector length	Bisection	M-section ($M = 33$)
5	2.38	1.86
15	3.10	1.98
30	3.54	2.17
40	3.74	2.45
50	3.88	2.60
100	4.34	2.99
200	4.76	3.03
300	5.02	3.06
400	5.20	3.10
500	5.32	3.13
1000	5.73	3.33

During the first iteration t_i is compared to x for $i = 1, 1 + L, 1 + 2L, \ldots,$ $1 + (M - 1)L$, where $M \leq 64$ and $1 + (M - 1)L \leq n$. In other words, M evenly spaced elements of T are compared with x. This iteration narrows the search to $L + 1$ elements, which are searched in the same way. The search repeats until $L = 1$, approximately $\log_m n$ iterations.

Although the M-section algorithm has a lot of overhead fetching the evenly spaced values from T, Dubois has shown that M-section is faster than bisection for all reasonable table sizes n. The results of Dubois' experiment are shown in Table 11-2.

11-8 ATTACHED PROCESSORS

An **attached processor** (often called an **array processor**, but not to be confused with a processor array) is a special-purpose pipelined processor designed to process large vectors or arrays. Being special-purpose computers, they must be attached to a general-purpose host computer. Attached processors are a lower-cost alternative to pipelined vector processors, generally offering a much better price-performance ratio.

Attached processors are similar to pipelined vector processors in several ways. They contain multiple pipelined functional units and parallel data paths. However, attached processors do not have vector instructions. Instead of vector instructions, attached processors achieve parallelism through the use of long instructions specifying many concurrent microoperations. A typical attached processor will have a large mathematical library of subroutines to perform array and matrix manipulations. This library of subroutines can be called from the host computer.

FPS-164/MAX

Not surprisingly, manufacturers of attached processors have taken advantage of parallelism to increase the power of their machines. A case in point is the FPS-164/MAX, marketed by Floating Point Systems. The FPS-164/MAX is an FPS-164 enhanced with up to 15 VLSI-based matrix-vector product modules (Figure 11-11). Each module contains two multiply-add pipelines. Since the FPS-164 contains one multiply-add pipeline, an FPS-164/MAX may contain up to 31 pipelines. Each pipeline has length 4. Hence the largest configuration machine can manipulate up to 124 vectors concurrently and has a peak performance rate of 330 megaflops. The first FPS-164/MAX was delivered in April 1985.

Matrix Multiplication

The MAX (Matrix Algebra Xcelerator) modules greatly improve the performance of the FPS-164 on linear algebra problems. As an example, let us consider computing the matrix product $C \leftarrow A \times B$. All the MAX modules are synchronized, and at any time in the algorithm every module is at exactly the same stage of computing eight matrix elements $c_{i,j}$. (Note that 8 equals 2 pipelines times 4 stages per pipeline.) See Figure 11-12.

FORTRAN code for this matrix multiplication algorithm is quite concise, emphasizing the power of the subroutine libraries. This program fragment is quoted from Charlesworth and Gustafson [1986].

```
C   MATRIX MULTIPLICATION (FPS-164/MAX):
    p = 2 * m + 1
    v = 4 * p
    DO 10 i = 1, n - v + 1, v
      CALL PLOAD (A(i,1), n, 1, n, v)
      DO 20 j = 1, n
        CALL PDOT (B(1,j), 1, n, C(i,j), 1, v)
20    CONTINUE
10  CONTINUE
```

To elaborate on the program, variable m is the number of MAX modules. Variable p is the number of pipelines. Recall that each MAX module has two pipelines and the FPS-164 has one. Each pipeline has four segments; variable v is the number of elements of **C** which will be computed concurrently. The outer **DO** loop marches through matrices **A** and **C**, v rows at a time. Subroutine **PLOAD** loads eight contiguous rows of matrix **A** into the vector registers of every MAX module and four contiguous rows into the table memory of the FPS-164. The inner **DO** loop computes the dot product n times, once for each column in **B**. Note that $n \times v$ dot products are performed for every row-loading operation. Subroutine **PDOT** zeros out the v scalar registers and then performs v dot product operations between the jth column of **B** and the v rows of **A** that are resident in the MAX modules and in the FPS-164 table memory.

Figure 11-13 illustrates the performance of this matrix multiplication algo-

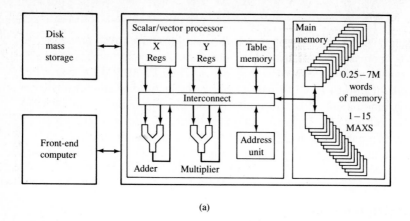

(a)

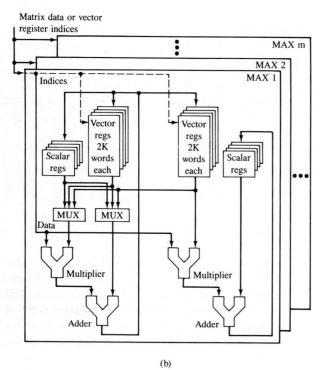

(b)

Figure 11-11 (a) FPS-164/MAX scientific computer, an attached processor. (b) MAX matrix-vector product module. (Reprinted by permission of Floating Point Systems.)

rithm. Note the familiar Amdahl effect: As the matrix dimension increases, so does peak performance. The "stair steps" in the performance curve are due to the fact that the total number of calls to **PDOT** is $n\lceil n/v \rceil$.

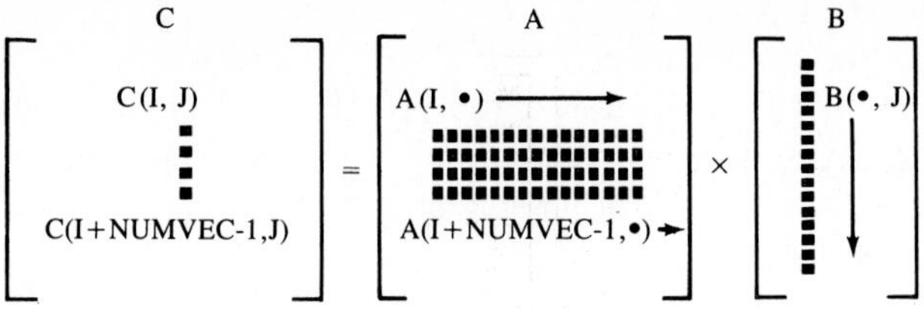

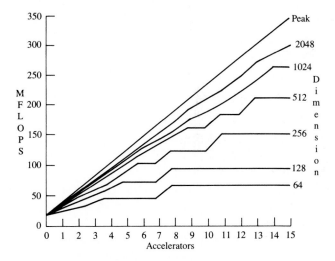

Figure 11-12 Matrix multiplication on the FPS-164/MAX. (Reprinted by permission of Floating Point Systems.)

Figure 11-13 Performance of matrix multiplication on the FPS-164/MAX. (Reprinted by permission of Floating Point Systems.)

11-9 SUMMARY

A number of pipelined vector processors have been commercial successes, having proved valuable tools for the processing of programs requiring extensive manipulations on vectors. Two noteworthy examples of commercial pipelined vector processors are the Cray-1 from Cray Research and the Cyber-205 from Control Data Corporation. The performance of these computers is dramatically increased when more than one pipelined vector operation can be chained together, providing an added measure of concurrency.

Hockney and Jesshope have defined a notation that allows the performance of pipelined vector processors on particular programs to be compared. Of course, different architectures excel on different programs.

Vectorizing compilers are becoming quite sophisticated at extracting vector instructions from blocks of sequential code. However, there is still a need for good algorithm design.

Attached processors are a low-cost alternative to pipelined vector processors and provide more performance per dollar. Unlike pipelined vector processors, attached processors do not have machine instructions that manipulate vectors. Rather, they rely upon carefully coded libraries of routines that use pipelining to achieve good performance on array and matrix manipulations. Attached processors may make use of substantial amounts of pipelining and parallelism in order to achieve their high performance.

BIBLIOGRAPHIC NOTES

Hwang and Briggs [1984] devote 180 pages to pipelining and vector processing. Many of the issues raised in this chapter are discussed in greater detail in their book. Hockney and Jesshope [1981] also discuss pipelined computers.

EXERCISES

11-1 Vector computers are constructed with a masking ability so that an instruction need not manipulate every element of a vector. Why is masking needed in a vector computer?

11-2 Given vector $\mathbf{A} = \{2, 3, -4, 6, 8\}$ and masking vector $\mathbf{M} = \{0, 1, 1, 0, 1\}$, what is the result of the compress instruction whose operands are $\mathbf{A}$ and $\mathbf{M}$?

11-3 Given vectors $\mathbf{A} = \{-3, 5, 3, 9\}$, $\mathbf{B} = \{2, 6, 10, 12\}$, and masking vector $\mathbf{M} = \{0, 1, 1, 1, 0, 0, 1, 0\}$, what is the result of the merge instruction whose operands are $\mathbf{A}$, $\mathbf{B}$, and $\mathbf{M}$?

11-4 In the context of pipelined vector processors, what is the difference between a register-to-register architecture and a memory-to-memory architecture?

11-5 Is low-order memory interleaving or high-order memory interleaving to be preferred on a pipelined vector processor? Why?

11-6 In the context of pipelined vector processors, what is strip mining?

11-7 The Cray-1 has smaller values of $n_{1/2}$ and r_∞ than the Cyber-205. Explain the ramifications of these measurements.

11-8 Acme's Bluejay-1 pipelined vector processor executes the following FORTRAN code segment

```
    DO 10 I = 1, N
       X(I) = Y(I) + Z(I)
   10 CONTINUE
```

in 2 microseconds when $\mathbf{N} = 10$, in 4.667 microseconds when $\mathbf{N} = 50$, and in 8 microseconds when $\mathbf{N} = 100$. Calculate r_∞ and $n_{1/2}$ for the Bluejay-1 executing this code.

11-9 The performance of the Cray-1 doing vector operations is severely degraded if every element of a vector to be loaded into a vector register is stored in the same memory bank. Under what circumstances would all the elements of a vector be stored in the same memory bank on the Cray-1?

11-10 The transmission lines carrying the clock signal from the CPU to the memory modules on the Cray-1 are all the same length. Why?

11-11 Changing from dyadic vector operations to linked triadic operations doubles the value of r_∞ on the Cyber-205. Why does the value of $n_{1/2}$ remain virtually unchanged?

11-12 What kind of performance could you expect from a vector computer executing an algorithm whose only vector manipulations were on sparse vectors (i.e., most elements to be ignored)?

11-13 Section 2-8 describes two problems to be solved on a Cray-1. Each subsection begins with a naïve algorithm and ends with a better algorithm. Which of the two optimizations described would a vectorizing compiler be more likely to discover? Why?

Acyclic—Refers to a graph without any cycles.

Adjacency Matrix—Boolean matrix indicating for each pair of vertices i and j whether there is an edge from i to j.

All-Pairs Shortest-Path Problem—Given a weighted graph, find the shortest path between every pair of vertices.

Alpha-Beta Pruning—An algorithm used to determine the minimax value of a game tree. Alpha-beta algorithm avoids searching subtrees whose evaluation cannot influence the outcome of the search.

Amdahl Effect—For any fixed number of processors, speedup is usually an increasing function of problem size.

AND Parallelism—Exploring conjunctive subtrees of an AND-OR tree in parallel (logic programming).

AND Tree—A search tree whose nonterminal nodes are all AND nodes.

AND/OR Tree—A search tree with both AND and OR nonterminal nodes.

Applicative Language—A language that performs all computations by applying functions to values.

Array Processor—(1) Attached processor; (2) processor array.

Associative Processor—A processor array with an associative memory.

Asynchronous Algorithm—See **Relaxation**.

Asynchronous Message Passing—A message passing scheme characterized by buffers of virtually unbounded length.

Attached Processor—A low-cost alternative to vector computer. An attached processor uses pipelining to manipulate vectors, but it does not have vector instructions.

AVL Tree—Binary tree having the property that for any node v in the tree, the difference in height between the left and right subtrees of node v is no more than 1.

Band matrix—This term is usually reserved for matrices in which the nonzero elements are clustered about the main diagonal.

Bandwidth—(1) Data transfer rate; (2) the value $1+2k$, where k is the largest number of columns any nonzero element of a matrix is from the main diagonal.

Bank—Unit of interleaved memory, allowing only a single read or write at a time.

Batch Search—A concurrent search for a number of names.

Binary Semaphore—A semaphore that can take on only the values 0 and 1.

Bitonic Merge—A parallel algorithm to merge two bitonic sequences of length 2^k into a single bitonic sequence of length 2^{k+1} in $k+1$ steps.

Bitonic Sequence—A sequence of numbers $a_0, a_1, \ldots, a_{n-1}$ with the property that (1) there exists an index i, $0 \leq i \leq n-1$, such that a_0 through a_i is monotonically increasing and a_i through a_{n-1} is monotonically decreasing, or else (2) there exists a cyclic shift of indices so that condition 1 is satisfied.

Blocking—A statement that sometimes delays the further execution of the invoking process.

Buffered Message Passing—A message passing scheme that uses buffers to store messages on their way from a sender to a receiver.

Busy-Waiting—Using processor cycles to test a variable until it assumes a desired value.

Butterfly Network—A processor organization containing $(k+1)2^k$ nodes, divided into 2^k columns and $k+1$ rows, or ranks.

Cache Memory—Small, fast memory unit used as a buffer between a processor and primary memory.

Channel—I/O processor.

Clausal Logic—A form of logic in which all propositions are expressed in terms of AND, OR, and NOT. A six-stage process transforms any predicate calculus formula into clausal form.

Clause—Sentence in clausal logic.

Cobegin/Coend—A structured way of indicating a set of statements that can be executed in parallel.

Communication Width—The size of shared memory in an SIMD model assuming a global memory.

Comparator—A device that performs the compare-exchange operation.

Compare-Exchange—Fundamental operation of bitonic merge algorithm. Two numbers are brought together, compared, and then exchanged, if necessary, so that they are in the proper order.

Complexity—Measure of time or space used by an algorithm. Without adjective, refers to time complexity.

Compress—A vector computer instruction that loads a vector according to the values in a corresponding masking vector.

Computer Module—Basic processor in Cm*.

Condition Synchronization—Delaying the continued execution of a process until some data object it shares with another process is in an appropriate state.

Connected—Given undirected graph G, the property that for every pair of vertices i and j in G, there is a path from i to j in G.

Control-Driven—Architecture with one or more program counters that determine the order in which instructions are executed.

Cost—Complexity of an algorithm multiplied by the number of processors used.

Critical Node—When a node is inserted into an AVL tree, the critical node is the root of the subtree about which a rebalancing is going to take place.

Critical Section—A sequence of statements that must appear to be executed as an atomic operation.

Cube-Connected Cycles Network—A processor organization that is a variant of a hypercube. Each hypercube node becomes a cycle of nodes, and no node has more than three connections to other nodes.

Cube-Connected Network—See **Hypercube**.

Cycle—Path, except that the first and last vertices are identical.

Cyclic Reduction—Parallel algorithm to solve general first-order linear recurrence relations.

Data-Driven—Data flow architecture in which execution of instructions depends on availability of operands.

Data Flow—An architecture in which the sequence of instruction execution depends not on a program counter, but on the availability of data.

Data Flow Analysis—Process of finding dependencies among instructions.

Data Flow Graph—(1) Machine language for a data flow computer; (2) result of data flow analysis.

Data Stream—Sequence of data used to execute an instruction stream.

Deadlock—A situation in which a set of active concurrent processes cannot proceed because each holds nonpreemptible resources that must be acquired by some other process.

Decision Problem—A problem whose solution, if any, is found by satisfying a set of constraints.

Dedicated Throughput—Number of results returned for a single job per time unit.

Demand-Driven—Data flow architecture in which execution of an instruction depends upon both availability of its operands and a request for the result.

Depth—Parallel time complexity.

Deque—A double-ended queue; i.e., a list of elements on which insertions and deletions can be performed at both the front and the rear.

Deterministic Model—A task model in which precedence relations between tasks and the execution time needed by each task are fixed and known before the schedule is devised.

Directed Graph—A graph in which the edges have an orientation, denoted by arrowheads.

Direct Naming—A message passing scheme in which source and destination designators are the names of processes.

Divide and Conquer—A problem-solving methodology that involves partitioning a problem into subproblems, solving the subproblems, and then combining the solutions to the subproblems into a solution for the original problem.

Dynamic Channel Naming—A message passing scheme allowing source and destination designators to be established at run time.

Dynamic Decomposition—A task allocation policy that assumes tasks are generated at execution time.

Edge—Component of a graph. An edge is a pair of vertices. If the edge is directed, the pair is ordered; if the edge is undirected, the pair has no order.

Efficiency—Ratio of speedup to number of processors used.

Enumeration Sort—A sort that finds the position of each name by determining the number of names smaller than it.

Expand—A vector computer instruction that stores the elements of a vector according to the values in a corresponding masking vector.

Expected Space Complexity—Average amount of space used by an algorithm over all possible inputs.

Expected Time Complexity—Average amount of time used by an algorithm over all possible inputs.

Fact—In the context of logic programming, a fact is a Horn clause with a head but no body.

Fork—A method of specifying concurrency. Similar to a procedure call, but the calling process continues execution. See **Join**.

From—In directed graph G, edge (u, v) is said to go from vertex u to vertex v.

Game Tree—State space tree representation of a game position.

Gantt Chart—Diagram used to illustrate a deterministic schedule.

Gather—Vector computer instruction that allows a level of indirection in determining elements of a vector to be read (loaded).

Gigaflops—Billion floating-point operations per second.

Global Name—See **Mailbox**.

Grain Size—Relative number of operations done between synchronizations in an MIMD algorithm.

Graph—Consists of V, a finite set of vertices, and E, a finite set of edges between pairs of vertices.

Height—In graph theory, the length of the longest path from the root of a tree to one of its leaves.

High-Order Interleaving—Memory interleaving based on high-order bits of an address.

Horizontal Processing—Processing a two-dimensional array row by row.

Horn Clause—A clause that contains at most one conclusion.

Hypercube—Boolean k-cube. Every processor has a unique address in the range from 0 to $2^k - 1$. Processors whose addresses differ in exactly 1 bit are connected. A hypercube network is a butterfly network with its columns collapsed into single nodes.

ICOT—Institute for New Generation Computer Technology in Japan.

Information—Collection of related data objects.

Inner-Product Method—Method of matrix multiplication in which one element of the resultant matrix is computed at a time.

Instruction Buffering—Prefetching instructions with the goal of never making the

CPU wait for an instruction to be fetched.

Instruction Look-Ahead—See **Instruction Buffering**.

Instruction Pipelining—Allowing more than one instruction to be in some stage of execution at the same time.

Instruction Stream—Sequence of instructions performed by a computer.

Interleaved Memory—Memory divided into a number of modules or banks that can be accessed simultaneously.

Internal Sort—Sorting a table of names contained entirely in primary memory.

Iterative Deepening—Use of a D-ply search to prepare for a $(D + 1)$-ply search.

Join—A method of synchronizing concurrent processes created by using a fork.

Kernel—The part of the operating system that schedules processes, among other duties.

Key—Unique object of a search.

Killer Heuristic—Heuristic suggesting that moves that have been found to refute other moves should be tried first (computer chess).

Knowledge—Information plus semantic meanings.

Knowledge Information Processing System—A computer system that processes knowledge, rather than data or information.

Linear Vector Scan—A table lookup algorithm for pipelined vector processors that in a single operation compares a large number of contiguous elements of the table against the key.

Linked Array—Data structure designed to allow the joining of various sized lists so that the inserting and deleting of the list elements can be done in parallel without contention.

Linked Triadic Operation—Executing a pair of vector operations, such as $V + S \times V$, as if they were a single, longer operation by taking the output of the first pipelined functional unit and directing it to the second pipelined functional unit, avoiding a trip to main memory and back (Cyber-205).

LIPS—See **Logical Inferences per Second**.

Locality of Reference—The observation that references to memory tend to cluster. Temporal locality refers to the observation that a particular datum or instruction, once referenced, is often referenced again in the near future. Spatial locality refers to the observation that once a particular location is referenced, a nearby location is often referenced in the near future.

Logarithmic Cost Criterion—Cost criterion that assumes the cost of performing an instruction is proportional to the length of the operands. The integer n requires $\lfloor \log n + 1 \rfloor$ bits of memory, hence the name.

Logic—The branch of mathematics that investigates the relationships between premises and conclusions of arguments.

Logical Inferences per Second—Procedure calls per second in a logic programming system.

Logic Program—A set of Horn clauses, or procedures.

Lower Triangular—A matrix with no nonzero elements above the main diagonal.

Low-Order Interleaving—Memory interleaving based on the low-order bits of the address.

LU Decomposition—Process of factoring a matrix **A** into a product of a lower triangular matrix **L** and an upper triangular matrix **U**.

M-Section—A table lookup algorithm for pipelined vector processors that combines features of bisection and linear vector scan.

Macropipelining—See **Pipelined Algorithm**.

Mailbox—An address used as a source or destination designator in a message.

Masking Vector—Used to enable and disable the processing of individual vector elements on a vector computer.

Megaflops—Million floating-point operations per second.

Memory Cycle Time—Time needed to access a memory location.

Mesh Network—A processor organization in which the nodes are arranged into a q-dimensional lattice and communication is allowed only between neighboring nodes.

Message-Oriented Language—A programming language in which process interaction is strictly through message passing.

Middle-Product Method—A method of matrix multiplication in which entire columns of the result are computed concurrently.

MIMD—Multiple-instruction stream, multiple data stream (Flynn). See **Multicomputer** and **Multiprocessor**.

Minimax—Algorithm used to determine the value of a game tree.

Minimum Spanning Tree—Given graph G, a spanning tree with the smallest possible weight among all spanning trees of G.

MISD—Multiple-instruction stream, single data stream (Flynn).

Module—Memory bank. Often used in context of interleaved memory.

Monitor—A structure consisting of variables representing the state of some resource, procedures to implement operations on that resource, and initialization code.

Multicomputer—A multiple-CPU computer designed for parallel processing, but lacking a shared memory.

Multiprocessor—A shared memory multiple-CPU computer designed for parallel processing.

Multiprogramming—Allowing more than one program to be in some state of execution at the same time.

Mutual Exclusion—The mutually exclusive execution of critical sections.

Necklace—The nodes a data item travels through in response to a sequence of shuffles.

Negamax—Knuth's implementation of the alpha-beta algorithm.

Nonblocking—A statement that never delays the execution of the invoking process.

Nondeterministic Model—A task model in which the execution time of each task is represented by a random variable.

Omega Network—A composition of shuffle-exchange networks with programmable switches.

Operation-Oriented Language—A programming language using remote procedure calls as the principal means for interprocess communication and synchronization.

Optimization Problem—A problem whose solution involves satisfying a set of constraints and minimizing (or maximizing) an objective function.

OR Parallelism—Exploring disjunctive subtrees of an AND-OR tree in parallel (logic programming).

OR Tree—A state space tree whose nonterminal nodes are all OR nodes.

Parallel Computation Thesis—Time on an unbounded parallel model of computation is (polynomially) equivalent to space on a sequential model of computation. (Unproved.)

Parallel Computer—A computer designed for the purpose of parallel processing.

Parallel Computing—The process of solving problems on parallel computers.

Parallelism—The use of multiple resources to increase concurrency.

Parallelizability—The ability to make an algorithm parallel.

Parallelization—Making an algorithm parallel.

Parallel Processing—A kind of information processing that emphasizes the concurrent manipulation of data elements belonging to one or more processes solving a single problem.

Parallel Unification—Finding the set of possible unifications for a goal in parallel.

Partial Cascade Sum—Parallel algorithm to compute partial sums in logarithmic time.

Partial Sum—Given a sequence $d_1, d_2, \ldots, d_n$ and an index i, the sum $\sum_{j=1}^{i} d_j$.

Partitioning—The division of data among processors in an MIMD algorithm.

Path—A sequence of edges $(v_1, v_2), (v_2, v_3), (v_3, v_4), \ldots, (v_{i-2}, v_{i-1}), (v_{i-1}, v_i)$ such that every vertex is in V, every edge is in E, and no two vertices are identical.

PDE—Partial differential equation.

Perfect Shuffle—A data routing scheme for 2^k nodes, numbered 0 through $2^k - 1$. A value at node i is routed to node $2i \bmod (2^k - 1)$, except the value at node $2^k - 1$ does not move to another node.

Pipelined Algorithm—In the context of multiple-CPU models, a pipelined algorithm is the software analog to pipelining in hardware. The algorithm is divided into an ordered set of segments in which the output of each segment is the input of its successor.

Pipelined Vector Processor—Vector computer implemented through the use of pipelined functional units. Examples include the Cray-1 and the Cyber-205.

Pipelining—Increasing concurrency by dividing a computation into a number of steps and allowing a number of tasks to be in various stages of execution at the same time.

Ports—A variant of mailboxes allowing multiple client processes but only a single server process.

Prescheduled—A kind of partitioned algorithm in which each processor is allocated its share of the computation at compile time.

Procedure—In the context of logic programming, a procedure is a Horn clause.

Procedure-Oriented Language—A programming language in which process communication and synchronization are accomplished through the use of shared variables.

Process Flow Graph—An acyclic directed graph in which vertices represent processes and edges represent execution constraints.

Processor Array—Vector computer implemented through the use of multiple processing elements. Examples include the ILLIAC IV and ICL DAP.

Prolog—Language for logic programming.

RAM—See **Random Access Machine**.

Random Access Machine—Simple model of a computer.

Random Uniform Game Tree—A game tree whose terminal node values are randomly chosen from some uniform distribution.

Rank—Row of a butterfly network.

Ready List—Operating system list containing ready-to-run processes.

Reasonable—A parallel model is reasonable if the number of processors each processor can communicate with directly is bounded by a constant.

Reduction—In Little et al.'s algorithm for the traveling salesperson problem, the process of determining the increase in the lower bound as the result of including or excluding an edge.

Refute—To make a move that causes an alpha-beta cutoff.

Relaxation—An algorithm design technique for MIMD models. Relaxed algorithms are characterized by the ability of processors to work with the most recently available data and the absence of synchronization points.

Remote Procedure Call—A structured implementation of a client-server interaction.

Rendezvous—When the server side of a remote procedure call is specified by using an `accept` statement or a similar construct.

Reply Message—Passes results back to the client in a remote procedure call.

Root—In Hirschberg's algorithm, the lowest-numbered vertex in a supervertex.

Rule—In the context of logic programming, a rule is a Horn clause with a head and a body.

Satisfiable—True under the rules of logic.

Scalable—An architecture is scalable if increasing the number of processors produces an analogous increase in the processing power of the machine.

Scatter—Vector computer instruction allowing one level of indirection in determining vector locations where another vector is to be written (stored).

Schedule—An allocation of tasks to processors.

Segment—Fundamental unit of a pipeline.

Self-scheduled—A kind of partitioned algorithm in which work is assigned to processes dynamically at run time.

Semaphore—Shared integer variable used for synchronization.

Send No-Wait—See **Asynchronous Message Passing**.

Sequential Algorithm—An algorithm designed to run on a sequential computer.

Sequential Computer—A computer whose instruction set includes operations on scalar variables only.

Serial Algorithm—See **Sequential algorithm**.

Serial Computer—See **Sequential computer**.

Short Necklace—A necklace of length less than k in a shuffle-exchange network containing 2^k nodes.

Shortstopping—Using the output of a functional unit before it is routed back to memory.

Shuffle-Exchange Network—Processor organization containing 2^k nodes, with two kinds of connections, called shuffle and exchange.

SIMD—Single-instruction stream, multiple data stream (Flynn).

SIMD-CC—Cube connected (hypercube) SIMD model.

SIMD-CCC—Cube-connected cycles SIMD model.

SIMD-MC—Mesh-connected SIMD model. Superscript refers to the number of dimensions in the mesh.

SIMD-P—SIMD model with pyramid processor organization.

SIMD-PS—SIMD model with perfect shuffle connections.

SIMD-SM—Shared-memory SIMD model. Two processors may not read from or write into the same memory location during the same instruction.

SIMD-SM-R—Shared-memory SIMD model. Two processors may read from the same memory location during an instruction, but concurrent writing into the same location is forbidden.

SIMD-SM-RW—Shared-memory SIMD model. Concurrent reading from and writing into the same memory location is allowed.

Single-Source Shortest-Path Problem—Problem of finding the shortest path from a single designated vertex (the source) to all the other vertices in a weighted, directed graph.

SISD—Single-instruction stream, single data stream (Flynn).

Software Lockout—Delay of processes due to contention for shared data resources.

Software Pipeline—The algorithmic-level equivalent to pipelining on the instruction level. An ordered set of algorithmic segments in which the output of each segment is the input of its successor.

SOR—Successive overrelaxation by points, a popular iterative method for solving partial differential equations on sequential computers.

Source—In the single-source shortest-path problem, the vertex from which all distances are calculated.

Space—Memory.

Space Complexity—Space used by an algorithm as a function of problem size.

Spanning Tree—Given a graph G, a tree that includes every vertex in G.

Speedup—Time taken to execute the best sequential algorithm solving problem Π divided by the time taken to execute a parallel algorithm solving Π.

Spin Lock—Shared variable used for busy waiting.

Spinning—A process waiting for the value of a spin lock to change.

Stage—See **Segment**.

State Space Tree—Representation of the decomposition of an original problem into subproblems through the addition of constraints.

Static Channel Naming—A message passing scheme in which source and destination designators are fixed at compile time.

Static Decomposition—Task allocation policy that assumes tasks and their precedence relations are known before execution.

Stream Parallelism—A pipelined variant of AND parallelism.

Stream Unit—Transmits vectors into the vector arithmetic section of the Cyber-205 CPU.

Strip Mining—In the context of a register-to-register pipelined vector processor, strip mining is the process of dividing a DO loop into a number of smaller loops that are executed sequentially, due to limitations in the length of vector registers.

Strongly Ordered Game Tree—Game tree with the following properties: (1) 70 percent of the time the first move chosen from any nonterminal node is the best move, and (2) 90 percent of the time the best move from any nonterminal node is one of the first 25 percent of the moves searched.

Strong Search—An algorithm that searches for a given key, locks the node associated with that key, and returns the node.

Subgraph—Given graph G, a graph whose vertices and edges are in G.

Supercomputer—A general-purpose computer capable of solving individual problems at extremely high computational speeds, compared with other computers built at the same time.

Superlinear Speedup—When the speedup exceeds the number of processors used.

Sword—A 512-bit superword on the Cyber-205.

Synchronized Algorithm—See **Partitioning**.

Synchronous Message Passing—Message passing without buffers in which both the send and receive processes are blocking.

Systolic Algorithm—In the context of multiple-CPU models, a systolic algorithm is the software analog to a systolic array. A special kind of pipelined algorithm.

Systolic Array—A collection of synchronized, special-purpose, rudimentary processors with a fixed interconnection network.

Table—Finite set of keys.

Throughput—Number of results produced per time unit.

Time—See **Time Complexity**.

Time Complexity—Time used by an algorithm as a function of problem size.

Timesharing—A kind of multiprogramming that allows a number of users to interact with their programs in real time.

To—In directed graph G, edge (u, v) is said to go from vertex u to vertex v.

Transposition Table—A hash table that stores previously evaluated game positions.

Tree—Connected, undirected, acyclic graph.

Turtle Program—Mapping notation used by Concurrent Prolog system on the Bagel.

Undirected Graph—A graph whose edges have no orientation.

Unification—Instantiation of a variable with a value.

Uniform Cost Criterion—Assumes every instruction takes 1 unit of time and every register requires 1 unit of space.

Upper Triangular—A matrix with no nonzero element below the main diagonal.

Utilization—Percentage of time a processor spends executing tasks.

Variable—Object whose identity is unspecified (logic programming).

Vector Computer—A computer with an instruction set that includes operations on vectors as well as scalars.

Vectorizing Compiler—A program that compiles high-level programs written in a sequential programming language, while converting as many blocks of sequential operations as possible into vector instructions.

Vector Looping—A mix of vertical and horizontal processing, vector looping processes a two-dimensional array by manipulating groups of rows, column by column.

Vertex—Component of a graph. Also called a *node*.

Vertical Processing—Processing a two-dimensional array column by column.

Weak Search—A search algorithm that searches for a key and returns the node that contained the key at the time it was examined. Weak search is not guaranteed to provide an up-to-date result.

Weight—Real number assigned to an edge in a weighted graph.

Weighted Graph—Graph with a real number assigned to each edge.

Weight Matrix—A matrix indicating, for each pair of vertices i and j, the weight of the edge from vertex i to vertex j.

Worst-Case Space Complexity—Greatest amount of space used by an algorithm over all possible inputs of a given size.

Worst-Case Time Complexity—Greatest amount of time used by an algorithm over all possible inputs of a given size.

BIBLIOGRAPHY

Ackerman, W. B. 1982. Data flow languages. *Computer* 14, 2 (February), pp. 15–25.

Agerwala, T., and Lint, B. 1978. Communication in parallel algorithms for boolean matrix multiplication. In *Proceedings of the 1978 International Conference on Parallel Processing* (August), pp. 146–153. IEEE, New York.

Aggarwal, A. 1984. A comparative study of X-tree, pyramid and related machines. *Proceedings of the 25th Annual Symposium on Foundations of Computer Science* (October), pp. 89–99. IEEE, New York.

Aho, A., Hopcroft, J., and Ullman, J. 1974. *The Design and Analysis of Computer Algorithms.* Addison-Wesley, Reading, MA.

Ahuja, N., and Swamy, S. 1984. Multiprocessor pyramid architecture for bottom-up image analysis. *IEEE Transactions on Pattern Analysis and Machine Intelligence*, PAMI-6, 4 (July), pp. 463–474.

Ahuja, S., Carriero, N., and Gelernter, D. 1986. Linda and friends. *Computer* 19, 8 (August), pp. 26–34.

Aigner, M. 1982. Parallel complexity of sorting problems. *Journal of Algorithms* 3, pp. 79–88.

Akl, S. G. 1982. A constant-time parallel algorithm for computing convex hulls. *BIT* 22, 2, pp. 130–134.

Akl, S. G. 1984a. An optimal algorithm for parallel selection. *Information Processing Letters* 19, 1, pp. 47–50.

Akl, S. G. 1984b. Optimal parallel algorithms for computing convex hulls and for sorting. *Computing* 33, 1, pp. 1–11.

Akl, S. G. 1985. *Parallel Sorting Algorithms.* Academic Press, Orlando, FL.

Akl, S. G., Barnard, D., and Doran, R. 1982. Design, analysis, and implementation of a parallel tree search algorithm. *IEEE Transactions on Pattern Analysis and Machine Intelligence*, PAMI-4 (March), pp. 192–203.

Akl, S. G., and Schmeck, H. 1984. Systolic sorting in a sequential input/output environment. In *Proceedings of the 22d Annual Allerton Conference on Communication, Control, and Computing* (October), pp. 946–955.

Alton, D. A., and Eckstein, D. M. 1979. Parallel breadth-first search of p-sparse graphs. In

Proceedings of the West Coast Conference on Combinatorics, Graph Theory and Computing (Humbolt State University, Arcata, California, September 5–7). In *Congressus Numerantium* 26.

Amdahl, G. 1967. Validity of the single processor approach to achieving large scale computing capabilities. In *AFIPS Conference Proceedings* 30 (April), pp. 483–485. Thompson Books, Washington, D.C.

Andrews, G. R. 1981. Synchronizing resources. *ACM Transactions on Programming Languages and Systems* 3, 4 (October), pp. 405–430.

Andrews, G. R. 1982. The distributed programming language SR—Mechanisms, design, and implementation. *Software Practice and Experience* 12, 8 (August), pp. 719–754.

Andrews, G. R., and Schneider, F. B. 1983. Concepts and notations for concurrent programming. *Computing Surveys* 15, 1 (March), pp. 3–43.

Arjomandi, E. 1975. A study of parallelism in graph theory. Ph.D. dissertation, University of Toronto, Toronto, Ontario, Canada.

Armstrong, P., and Rem, M. 1982. A serial sorting machine. *Computers and Electrical Engineering* 9, 1, pp. 53–58.

Atallah, M. J. 1983. Algorithms for VLSI networks of processors. Ph.D. dissertation, The Johns Hopkins University, Baltimore, MD.

Atallah, M. J., and Kosaraju, S. R. 1982. Graph problems on a mesh-connected processor array (preliminary version). In *Proceedings of the 14th Annual ACM Symposium on Theory of Computing* (May), pp. 345–353. ACM, New York.

Atallah, M. J., and Kosaraju, S. R. 1984. Graph problems on a mesh-connected processor array. *Journal of the ACM* 31, 3 (July), pp. 649–667.

Atjai, M., Komlós, J., and Szemerédi, E. 1983. An $O(n \log n)$ sorting network. In *Proceedings of the 15th Annual ACM Symposium on the Theory of Computing* (May), pp. 1–9.

Baase, S. 1978. *Computer Algorithms: Introduction to Design and Analysis*. Addison-Wesley, Reading, MA.

Baer, J.-L. 1980. *Computer Systems Architecture*. Computer Science Press, Potomac, MD.

Baer, J.-L. 1982. Techniques to exploit parallelism. In *Parallel Processing Systems: An Advanced Course*, D. J. Evans, ed. Cambridge University Press, Cambridge, England, pp. 75–99.

Baer, J.-L., Du, H.-C., and Ladner, R. E. 1983. Binary search in a multiprocessing environment. *IEEE Transactions on Computers* C-32, 7 (July), pp. 667–677.

Baer, J.-L., and Schwab, B. 1977. A comparison of tree-balancing algorithms. *Communications of the ACM* 20, 5 (May), pp. 322–330.

Balzer, R. M. 1971. PORTS—A method for dynamic interprogram communication and job control. In *Proceedings of the AFIPS Spring Joint Computer Conference* (Atlantic City, NJ, May 18–20), vol. 38. AFIPS Press, Arlington, VA, pp. 485–489.

Barnes, G. H., Brown, R. M., Kato, M., Kuck, D. J., Slotnick, D. L., and Stokes, R. A. 1968. The Illiac IV computer. *IEEE Transactions on Computers* C-27, 1 (January), pp. 84–87.

Batcher, K. E. 1968. Sorting networks and their applications. In *Proceedings of the Spring Joint Computer Conference* (Atlantic City, NJ, Apr. 30–May 2), vol. 32. AFIPS Press, Reston, VA, pp. 307–314.

Batcher, K. E. 1979. The STARAN Computer. In *Infotech State of the Art Report: Supercomputers*, vol. 2, C. R. Jesshope and R. C. Hockney, eds. Infotech, Maidenhead, England, pp. 33–49.

Batcher, K. E. 1980. Design of massively parallel processor. *IEEE Transactions on Computers* C-29, pp. 836–840.

Baudet, G. M. 1978a. Asynchronous iterative methods for multiprocessors. *Journal of the ACM* 25, 2 (April), pp. 226–244.

Baudet, G. M. 1978b. The design and analysis of algorithms for asynchronous multiprocessors. Ph.D. dissertation, Carnegie-Mellon University, Pittsburgh.

Baudet, G., and Stevenson, D. 1978. Optimal sorting algorithms for parallel computers. *IEEE Transactions on Computers* C-27, 1 (January), pp. 84–87.

BBN. 1985. Butterfly^{TM} parallel processor overview. Version 1. Tech. Rept., BBN Laboratories, Inc. (December 18). Cambridge, MA.

Ben-Ari, M. 1982. *Principles of Concurrent Programming*. Prentice-Hall, Englewood Cliffs, NJ.

Bentley, J. L. 1980. A parallel algorithm for constructing minimum spanning trees. *Journal of Algorithms* 1, 1 (March), pp. 51–59.

Bentley, J. L., and Brown, D. J. 1980. A general class of recurrence tradeoffs. In *Proceedings of the 21st Annual Symposium on Foundations of Computer Science* (October), pp. 217–228. IEEE, New York.

Bentley, J. L. and Kung, H. T. 1979. A tree machine for searching problems. In *Proceedings of the 1979 International Conference on Parallel Processing* (August), pp. 257–266. IEEE, New York.

Berg, H. K., Boebert, W. E., Franta, W. R., and Moher, T. G. 1982. *Formal Methods of Program Verification and Specification*. Prentice-Hall, Englewood Cliffs, NJ, chap. 6.

Bergland, G. D. 1972. A parallel implementation of the Fast Fourier Transform algorithm. *IEEE Transactions on Computers* C-21, 4 (April), pp. 366–370.

Berliner, H., and Ebeling, C. 1986. The SUPREM architecture: A new intelligent paradigm. *Artificial Intelligence* 28, pp. 3–8.

Bernhard, R. 1982. Computing at the speed limit. *IEEE Spectrum* 19, 7, pp. 26–31.

Bhatt, S. N., and Leiserson, C. E. 1982. How to assemble tree machines. *Proceedings of the 14th Annual ACM Symposium on Theory of Computing* (May), pp. 77–84. ACM, New York.

Bilardi, G., Pracchi, M., and Preparata, F. P. 1981. A critique and an appraisal of VLSI models of computation. In *VLSI Systems and Computations*, H. T. Kung, B. Sproull, and G. Steele, eds. Springer-Verlag, New York, pp. 81–88.

Bilardi, G., and Preparata, F. P. 1983. A VLSI optimal architecture for bitonic sorting. In *Proceedings of the 7th Conference on Information Science Systems*, pp. 1–5.

Bilardi, G., and Preparata, F. P. 1984a. A minimum area VLSI architecture for $O(\log n)$ time sorting. In *Proceedings of the 16th Annual ACM Symposium on Theory of Computing* (May), pp. 64–70. ACM, New York.

Bilardi, G., and Preparata, F. P. 1984b. An architecture for bitonic sorting with optimal VLSI performance. *IEEE Transactions on Computers* C-33, 7 (July), pp. 646–651.

Bitton, D., DeWitt, D. J., Hsaio, D. K., and Menon, J. 1984. A taxonomy of parallel sorting. *ACM Computing Surveys* 16, 3 (September), pp. 287–318.

Bitton-Friedland, D. 1982. Design, analysis and implementation of parallel external sorting algorithms. Ph.D. dissertation, University of Wisconsin-Madison.

Bonuccelli, M. A., Lodi, E., and Pagli, L. 1984. External sorting in VLSI. *IEEE Transactions on Computers* C-33, 10 (October), pp. 931–934.

Borodin, A., and Hopcroft, J. E. 1982. Routing, merging, and sorting on parallel models of computation. In *Proceedings of the 14th Annual ACM Symposium on Theory of Computing* (May), pp. 338–344. ACM, New York.

Brandenburg, J. E., and Scott, D. S. 1985. Embeddings of communication trees and grids into hypercubes. Tech. Rept., Intel Scientific Computers.

Brigham, E. O. 1973. *The Fast Fourier Transform*. Prentice-Hall, Englewood Cliffs, NJ.

Brinch Hansen. 1973a. *Operating System Principles*. Prentice-Hall, Englewood Cliffs, NJ.

Brinch Hansen. 1973b. Concurrent programming concepts. *ACM Computing Surveys* 5, 4 (December), pp. 223–245.

Brinch Hansen, P. 1975. The programming language Concurrent Pascal. *IEEE Transactions on Software Engineering* SE-1, 2 (June), pp. 199–206.

Brinch Hansen, P. 1977. *The Architecture of Concurrent Programs*. Prentice-Hall, Englewood Cliffs, NJ.

Brinch Hansen, P. 1978. Distributed processes: A concurrent programming concept. *Communications of the ACM* 21, 11 (November), pp. 934–941.

Brinch Hansen, P. 1981. Edison: A multiprocessor language. *Software Practice and Experience*

11, 4 (April), pp. 325–361.

Brock, H. K., Brooks, B. J., and Sullivan, F. 1981. Diamond: A sorting method for vector machines. *BIT* 21, pp. 142–152.

Browning, S. A. 1980a. The tree machine: A highly concurrent computing environment. Ph.D. dissertation, California Institute of Technology, Pasadena.

Browning, S. A. 1980b. Algorithms for the tree machine. In *Introduction to VLSI Systems*, C. Mead and L. Conway, eds. Addison-Wesley, Reading, MA.

Bucher, I. Y., and Simmons, M. L. 1985. Performance assessment of supercomputers. Preprint LA-UR-85-1505, Los Alamos National Laboratory, Los Alamos, NM. (To appear in *Vector and Parallel Processors: Architecture, Applications, and Performance Evaluation*, M. Ginsberg, ed., North-Holland.)

Burton, F. W., and Huntbach, M. M. 1984. Virtual tree machines. *IEEE Transactions on Computers* C-33, 3 (March), pp. 278–280.

Carey, M. J., Hansen, P. M., and Thompson, C. D. 1982. RESST: A VLSI implementation of a record-sorting stack. Tech. Rept. UCB/CSD 82/102, Computer Science Division, University of California, Berkeley.

Carey, M. J., and Thompson, C. D. 1984. An efficient implementation of search trees on $\lceil \lg N + 1 \rceil$ processors. *IEEE Transactions on Computers* C-33, 11 (November), pp. 1038–1041.

Chabbar, E. 1980. Contrôle et gestion du parallélisme: tris synchrones et asynchrones. Thesis, Université de Franche-Comté, France.

Chandra, A. K. 1976. Maximal parallelism in matrix multiplication. IBM Tech. Rept. RC6193, Thomas J. Watson Research Center, Yorktown Heights, NY.

Chandra, A. K., and Stockmeyer, L. J. 1976. Alternation. In *Proceedings of the 17th Annual Symposium on Foundations of Computer Science* (October), pp. 98–108. IEEE, New York.

Chang, S.-K. 1974. Parallel balancing of binary search trees. *IEEE Transactions on Computers* C-23, 4 (April), pp. 441–445.

Charlesworth, A. E., and Gustafson, J. L. 1986. Introducing replicated VLSI to supercomputing: the FPS-164/MAX scientific computer. *Computer* 19, 3 (March), pp. 10-22.

Chazelle, B. M., and Monier, L. M. 1981a. A model of computation for VLSI with related complexity results. In *Proceedings of the 13th Annual ACM Symposium on Theory of Computing* (May), pp. 318–325. ACM, New York.

Chazelle, B. M., and Monier, L. M. 1981b. Optimality in VLSI. In *VLSI 81*, J.P. Gray, ed. Academic Press, London, pp. 269–278.

Chen, A. C., and Wu, C.-L. 1984. Optimum solution to dense linear systems of equations. In *Proceedings of the 1984 International Conference on Parallel Processing* (August), pp. 417–424. IEEE, New York.

Chen, I.-N. 1975. A new parallel algorithm for network flow problems. In *Parallel Processing, Lecture Notes in Computer Science*, vol. 24. Springer-Verlag, New York, pp. 306–307.

Chen, T. C., Eswaran, K. P., Lum, V. Y., and Tung, C. 1978a. Simplified odd-even sort using multiple shift-register loops. *International Journal of Computer and Information Science* 7, 3, pp. 295–314.

Chen, T. C., Lum, V. Y., and Tung, C. 1978b. The rebound sorter: An efficient sort engine for large files. In *Proceedings of the 4th International Conference on Very Large Data Bases*, West Berlin, Germany, pp. 312–318.

Chen, Y. K., and Feng, T. 1973. A parallel algorithm for maximum flow problem. In *Proceedings of the 1973 Computer Conference on Parallel Processing* (Sagamore, NY) (August), p. 60.

Cheriton, D., and Tarjan, R.E. 1976. Finding minimum spanning trees. *SIAM Journal on Computing* 5, 4 (December), pp. 724–742.

Chern, M.-Y., and Murata, T. 1983a. A fast algorithm for concurrent LU decomposition and matrix inversions. In *Proceedings of the 1983 International Conference on Parallel Processing* (August), pp. 79–86. IEEE, New York.

Chern, M.-Y., and Murata, T. 1983b. Efficient matrix multiplications on a concurrent data-loading array processor. In *Proceedings of the 1983 International Conference on Parallel Processing* (August), pp. 90–94. IEEE, New York.

Cheung, J., Dhall, S., Lakshmivarahan, S., Miller, L., and Walker, B. 1982. A new class of two stage parallel sorting schemes. In *Proceedings of the ACM '82 Conference* (October), pp. 26–29. ACM, New York.

Chin, F. I., and Fok, K. S. 1980. Fast sorting algorithms on uniform ladders (multiple shift register loops). *IEEE Transactions on Computers* C-29, 7 (July), pp. 618–631.

Chin, F. Y., Lam, J., and Chen, I.-N. 1981. Optimal parallel algorithms for the connected component problem. In *Proceedings of the 1981 International Conference on Parallel Processing* (August), pp. 170–175. IEEE, New York.

Chin, F. Y., Lam, J., and Chen, I.-N. 1982. Efficient parallel algorithms for some graph problems. *Communications of the ACM* 25, 9 (September), pp. 659–665.

Chow, P., Vranesic, Z. G., and Yen, J. L. 1983. A pipeline distributed arithemetic PFFT processor. *IEEE Transactions on Computers* C-32, 12 (December), pp. 1128–1136.

Chung, K.-M., Luccio, F., and Wong, C. K. 1980a. On the complexity of sorting on magnetic bubble memory systems. *IEEE Transactions on Computers* C-29, 7 (July), pp. 553–563.

Chung, K.-M., Luccio, F., and Wong, C. K. 1980b. Magnetic bubble memory structures for efficient sorting and searching. In *Proceedings IFIP Congress: Information Processing 80*, pp. 439–444.

Cichelli, R. J. 1973. Research progress report in computer chess. *SIGART Newsletter* 41 (August), pp. 32–36.

Clocksin, W. F., and Mellish, C. S. 1981. *Programming in Prolog*. Springer-Verlag, New York.

Coffman, E. G., Jr., and Denning, P. J. 1973. *Operating Systems Theory*. Prentice-Hall, Englewood Cliffs, NJ.

Cole, R. 1984. Slowing down sorting networks to obtain faster sorting algorithms. In *Proceedings of the 25th Annual Symposium on Foundations of Computer Science* (October), pp. 255–260. IEEE, New York.

Colmerauer, A., Kanoui, H., Pasero, R., and Roussel, P. 1973. Un système de comunication homme-machine en Français. Rapport, Groupe Intelligence Artificielle, Universite d'Aix Marseille, Luminy.

Conery, J., and Kibler, D. 1984. AND parallelism in logic programming. Unpublished. *Tutorial on Parallel Logic Programming, International Conference on Parallel Processing* (August), pp. 13–17.

Conway, M. E. 1963. A multiprocessor system design. In *Proceedings AFIPS Fall Joint Computer Conference* (Las Vegas, NV, November), vol. 24. Spartan Books, Baltimore, MD, pp. 139–146.

Cook, S. A. 1974. An observation on time-storage trade-off. *Journal of Computer and System Sciences* 9, 3, pp. 308–316.

Corinthios, M. J., and Smith, K. C. 1975. A parallel radix-4 fast Fourier transform computer. *IEEE Transactions on Computers* C-24, 1 (January), pp. 80–92.

Crane, B. A. 1968. Path finding with associative memory. *IEEE Transactions on Computers* C-17, 7 (July), pp. 691–693.

Crane, B. A., Gilmartin, M. J., Huttenhoff, J. H., Rux, P. T., and Shively, R. R. 1972. PEPE computer architecture. *COMPCON 72 Digest*, IEEE, New York, pp. 57–60.

Cyre, W. R., and Lipovski, G. J. 1972. On generating multipliers for a cellular fast Fourier transform processor. *IEEE Transactions on Computers* C-21, 1 (January), pp. 83–87.

Davis, A. L., and Keller, R. M. 1982. Data flow program graphs. *Computer* 14, 2 (February), pp. 26–41.

De Bruijn, N. G. 1984. Some machines defined by directed graphs. *Theoretical Computer Science* 32, pp. 309–319.

deGroot, A. D. 1965. *Thought and Choice in Chess*. Mouton, The Hague.

Dekel, E., Nassimi, D., and Sahni, S. 1981. Parallel matrix and graph algorithms. *SIAM Journal on Computing* 10, 4 (November), pp. 657–675.

Dekel, E., and Sahni, S. 1982. A parallel matching algorithm for convex bipartite graphs. In *Proceedings of the 1982 International Conference on Parallel Processing* (August), pp. 178–184. IEEE, New York.

Deminet, J. 1982. Experience with multiprocessor algorithms. *IEEE. Transactions on Computers* C-31, 4 (April), pp. 278–288.

Demuth, H. B. 1956. Electronic data sorting. Ph.D. dissertation, Stanford University, Stanford, CA.

Dennis, J. B., and Van Horn, E. C. 1966. Programming semantics for multiprogrammed computations. *Communications of the ACM* 9, 3 (March), pp. 143–155.

Deo, N. 1974. *Graph Theory with Applications to Engineering and Computer Science.* Prentice-Hall, Englewood Cliffs, NJ.

Deo, N., Pang, C. Y., and Lord, R. E. 1980. Two parallel algorithms for shortest path problems. In *Proceedings of the 1980 International Conference on Parallel Processing* (August). pp. 244–253. IEEE, New York.

Deo, N., and Yoo, Y. B. 1981. Parallel algorithms for the minimum spanning tree problem. In *Proceedings of the 1981 International Conference on Parallel Processing* (August), pp. 188–189. IEEE, New York.

Dere, W. Y., and Sakrison, D. J. 1970. Berkeley array processor. *IEEE Transactions on Computers* C-19, 5 (May), pp. 444–447.

Desai, B. C. 1978. The BPU: A staged parallel processing system to solve the zero-one problem. In *Proceedings ICS '78* (December), pp. 802–817.

Dijkstra, E. W. 1959. A note on two problems in connexion with graphs. *Numererische Mathematik* 1, pp. 269–271.

Dijkstra, E. W. 1968a. The structure of the 'THE' multiprogramming system. *Communications of the ACM* 11, 5 (May), pp. 341–346.

Dijkstra, E. W. 1968b. Cooperating sequential processes. In *Programming Languages*, F. Genuys, ed. Academic Press, New York.

Douglass, R. J. 1984. Characterizing the parallelism in rule-based expert systems. Tech. Rept. LA-UR-84-3428, Los Alamos National Laboratory, Los Alamos, NM.

Dowd, M., Perl, Y., Rudolph, L., and Saks, M. 1983. The balanced sort network. In *Proceedings of the Conference on Principles of Distributed Computing*, pp. 161–172.

Dubois, P. F. 1982. Swimming upstream: Calculating table lookups and piecewise functions. In *Parallel Computations*, G. Rodrigue, ed. Academic Press, New York, pp. 129–151.

Dyer, C. R. 1981. A VLSI pyramid machine for hierarchical parallel image processing. In *Proceedings PRIP*, pp. 381–386.

Dyer, C. R. 1982. Pyramid algorithms and machines. In *Multicomputers and Image Processing Algorithms and Programs*, K. Preston and L. Uhr, eds. Academic Press, New York, pp. 409–420.

Dyer, C. R., and Rosenfeld, A. 1981. Parallel image processing by memory augmented cellular automata. *IEEE Transactions on Pattern Analysis and Machine Intelligence*, PAMI-3, pp. 29–41.

Eckstein, D. M. 1979. BFS and biconnectivity. Tech. Rept. 79-11, Dept. of Computer Science, Iowa State University of Science and Technology, Ames.

Eckstein, D. M., and Alton, D. A. 1977a. Parallel searching of non-sparse graphs. Tech. Rept. 77-02, Dept. of Computer Science, The University of Iowa, Iowa City.

Eckstein, D. M., and Alton, D. A. 1977b. Parallel graph processing using depth-first search. In *Proceedings of the Conference on Theoretical Computer Science* (Waterloo, Ontario, August), University of Waterloo, Waterloo, Ontario, pp. 21–29.

Ein-Dor, P. 1985. Grosch's law re-revisited: CPU power and the cost of computation. *Communications of the ACM* 28, 2 (February), pp. 142–151.

El-Dessouki, O. I., and Huen, W. H. 1980. Distributed enumeration on network computers. *IEEE Transactions on Computers* C-29, 9 (September), pp. 818–825.

Ellis, C. 1980a. Concurrent search and insertion in 2-3 trees. *Acta Informatica* 14 (1980), pp. 63–86.

Ellis, C. 1980b. Concurrent search and insertion in AVL trees. *IEEE Transactions on Computers* C-29, 9 (September), pp. 811–817.

Enslow, P. H., ed. 1974. *Multiprocessors and Parallel Processing.* Wiley, New York.

Enslow, P. H. 1977. Multiprocessor organization—A survey. *Computing Surveys* 9, 1 (March), pp. 102–129.

Enslow, P. H. 1978. What is a "distributed" processing system? *IEEE Computer* 11, 1 (January), pp. 13–21.

Eppinger, J. L. 1983. An empirical study of insertion and deletion in binary search trees. *Communications of the ACM* 26, 9 (September), pp. 663–669.

ETA. 1984. Kuck and associates to develop advanced preprocessor for ETA[10]. *ETA Systems i/o* 1, 2 (Fall 1984), p. 2.

Even, S. 1974. Parallelism in tape-sorting. *Communications of the ACM* 17, 4 (April), pp. 202–204.

Falk, H. 1976. Reaching for the gigaflop. *IEEE Spectrum* 13, 10 (October), pp. 64–70.

Feierbach, G., and Stevenson, D. 1979. The ILLIAC IV. In *Infotech State of the Art Report: Supercomputers*, vol. 2, C. R. Jesshope and R. W. Hockney, eds. Infotech, Maidenhead, England, pp. 77–92.

Feldman, J. A. 1979. High level programming for distributed computing. *Communications of the ACM* 22, 6 (June), pp. 353–368.

Feng, T.-Y. 1981. A survey of interconnection networks. *Computer* 14, 12 (December).

Finkel, R., and Fishburn, J. 1982. Parallelism in alpha-beta. *Artificial Intelligence*, pp. 89–106.

Fishburn, J. P. 1981. Analysis of speedup in distributed algorithms. Ph.D. dissertation, University of Wisconsin-Madison.

Fishburn, J. P., and Finkel, R. A. 1982. Quotient networks. *IEEE Transactions on Computers* C-31, 4 (April).

Flanders, P. M. 1982. A unified approach to a class of data movements on an array processor. *IEEE Transactions on Computers* C-31, 9 (September), pp. 809–819.

Flanders, P. M., Hunt, D.J., Reddaway, S. F., and Parkinson, D. 1977. Efficient high speed computing with the Distributed Array Processor. In *High Speed Computer and Algorithm Organisation*. Academic Press, London, pp. 113–128.

Flanders, P. M., and Reddaway, S. F. 1984. Sorting on DAP. In *Parallel Computing 83*, M. Feilmeier, G. Joubert, and U. Schendel, eds. North-Holland, Amsterdam, pp. 247–252.

Floyd, R. W. 1962. Algorithm 97: Shortest path. *Communications of the ACM* 5, 6 (June), p. 345.

Flynn, M. J. 1966. Very high-speed computing systems. *Proceedings of the IEEE* 54, 12 (December), pp. 1901–1909.

Foster, C. C. 1976. *Content-Addressable Parallel Processors.* Van Nostrand Reinhold, New York.

Foster, M. J., and Kung, H. T. 1980. Design of special-purpose VLSI chips—Example and opinions. *Computer* 13, 1 (January), pp. 26–40.

Frenkel, K. A. 1986. Evaluating two massively parallel machines. *Communications of the ACM* 29, 8 (August), pp. 752–758.

Fritsch, G., Kleinoeder, W., Linster, C. U., and Volkert, J. 1983. EMSY85—The Erlanger multi-processor system for a broach spectrum of applications. In *Proceedings of the 1983 International Conference on Parallel Processing* (August), pp. 325–330. IEEE, New York.

Fuller, S. H., and Oleinick, P. N. 1976. Initial measurements of parallel programs in a multi-miniprocessor. In *Proceedings of the 13th IEEE Computer Society International Conference*, IEEE, New York, pp. 358–363.

Fung, L. 1977. A massively parallel processing computer. In *High Speed Computer and Algorithm Organisation*, D. J. Kuck, D. H. Lawrie, and A. H. Sameh, eds. Academic Press, London, pp. 203–204.

Gajski, D. D., Padua, D. A., Kuck, D. J., and Kuhn, R. H. 1982. A second opinion on data flow machines and languages. *Computer* (February), pp. 58–69.

Galil, Z. 1976. Hierarchies of complete problems. *Acta Informatica* 6, pp. 77–88.

Galil, Z., and Paul, W. J. 1981. An efficient general purpose parallel computer. In *Proceedings of the 13th Annual ACM Symposium on Theory of Computing* (May), pp. 247–262. ACM, New York.

Galil, Z., and Paul, W. J. 1983. An efficient general-purpose parallel computer. *Journal of the ACM* 30, 2 (April), pp. 360–387.

Garey, M. R., and Johnson, D. S. 1979. *Computers and Intractability: A Guide to the Theory of NP-Completeness.* W. H. Freeman, San Francisco.

Gavril, F. 1975. Merging with parallel processors. *Communications of the ACM* 18, 10 (October), pp. 588–591.

Gehringer, E. F., Jones, A. K., and Segall, Z. Z. 1982. The Cm* testbed. *Computer* (October), pp. 40–53.

Gentleman, W. M. 1978. Some complexity results for matrix computations on parallel computers. *Journal of the ACM* 25, 1 (January), pp. 112–115.

Gilbert, E. J. 1983. Algorithm partition tools for a high-performance multiprocessor. Ph.D. dissertation, Stanford University, Stanford, CA.

Gillogly, J. 1978. Performance analysis of the technology chess program. Ph.D. dissertation, Carnegie-Mellon University, Pittsburgh.

Gilmore, P. A. 1974. Matrix computations on an associative processor. In *Lecture Notes in Computer Science*, vol. 24, Parallel Processing. Springer-Verlag, New York, pp. 272–290.

Golden, B., Bodin, L., Doyle, T., and Stewart, W., Jr. 1980. Approximate traveling salesman algorithms. *Operations Research* 28, 3 (May–June), Part 2, pp. 694–711.

Goldschlager, L. M. 1977. The monotone and planar circuit value problems are log space complete for *P*. *SIGACT News* 9, 2 (Summer), pp. 25–29.

Goldschlager, L. M. 1978. A unified approach to models of synchronous parallel machines. In *Proceedings of the 10th Annual ACM Symposium on Theory of Computing* (May), pp. 89–94. ACM, New York.

Goldschlager, L. M. 1982. A universal interconnection pattern for parallel computers. *Journal of the ACM* 29, 4 (October), pp. 1073–1086.

Goldschlager, L. M., Shaw, R. A., and Staples, J. 1982. The maximum flow problem is log space complete for *P*. *Theoretical Computer Science* 21 (October), pp. 105–111.

Goodman, S. E., and Hedetniemi, S. T. 1977. *Introduction to the Design and Analysis of Algorithms.* McGraw-Hill, New York, p. 265.

Gottlieb, A. 1986. An overview of the NYU Ultracomputer project. Ultracomputer Note 100, Ultracomputer Research Laboratory, Division of Computer Science, Courant Institute of Mathematical Sciences, New York University, New York.

Gottlieb, A., Grishman, R., Kruskal, C. P., McAuliffe, K. P., Rudolph, L., and Snir, M. 1983. The NYU ultracomputer: Designing an MIMD shared memory parallel computer. *IEEE Transactions on Computers* C-32, 2, pp. 175–189.

Graham, R. L. 1972. Bounds on multiprocessing anomalies and packing algorithms. In *Proceedings AFIPS 1972 Spring Joint Computer Conference*, pp. 205–217.

Greif, I. 1977. A language for formal problem specifications. *Communications of the ACM* 20, 12 (December), pp. 931–935.

Grosch, H. A. 1953. High speed arithmetic: The digital computer as a research tool. *Journal of the Optical Society of America* 43, 4 (April).

Grosch, H. A. 1975. Grosch's law revisited. *Computerworld* 8, 16 (April 16), p. 24.

Guibas, L. J., Kung, H. T., and Thompson, C. D. 1979. Direct VLSI implementation of combinatorial problems. In *Proceedings of the Conference on Very Large Scale Integration: Architecture, Design, Fabrication* (Pasadena, CA, January), California Institute of Technology, Pasadena, pp. 509–525.

Gurd, J. R., Kirkham, C. C., and Watson, I. 1985. The Manchester prototype dataflow computer. *Communications of the ACM* 28, 1 (January), pp. 34–52.

Habermann, A. N. 1972. Parallel neighbor sort. Tech. Rept., Carnegie-Mellon University, Pittsburgh.

Häggvist, R., and Hell, P. 1981. Parallel sorting with constant time for comparisons. *SIAM Journal on Computing* 10, 3 (August), pp. 465–472.

Halstead, R. H., Jr. 1986. Parallel symbolic computing. *Computer* 19, 8 (August), pp. 35–43.

Hambrusch, S. E. 1982. The complexity of graph problems on VLSI. Ph.D. dissertation, The Pennsylvania State University, University Park.

Händler, W. 1977. The impact of classification schemes on computer architecture. *Proceedings of the 1977 International Conference on Parallel Processing* (August), pp. 7–15. IEEE, New York.

Harary, F. 1969. *Graph Theory.* Addison-Wesley, Reading, MA.

Harrison, T. J., and Wilson, M. W. 1983. Special-purpose computers. In *Encyclopedia of Computer Science and Engineering*, 2d ed. A. Ralston, ed. Van Nostrand-Reinhold, New York, pp. 1385–1393.

Hayes, J. P. 1978. *Computer Architecture and Organization.* McGraw-Hill, New York.

Haynes, L. S., Lau, R. L., Siewiorek, D. P., and Mizell, D. 1982. A survey of highly parallel computing. *Computer* 14, 1 (January), pp. 9–24.

Heller, D. 1978. A survey of parallel algorithms in numerical linear algebra. *SIAM Review* 20 (1978), pp. 740–777.

Helman, P., and Veroff, R. 1986. *Intermediate Problem Solving and Data Structures: Walls and Mirrors.* Benjamin/Cummings, Menlo Park, CA.

Higbie, L. C. 1972. The OMEN computers: Associative array processors. In *COMPCON 72 Digest.* IEEE, New York, pp. 287–290.

Hirschberg, D. S. 1976. Parallel algorithms for the transitive closure and the connected component problem. In *Proceedings of the 8th Annual ACM Symposium on the Theory of Computing* (May), pp. 55–57. ACM, New York.

Hirschberg, D. S. 1978. Fast parallel sorting algorithms. *Communications of the ACM* 21, 8 (August), pp. 657–666.

Hirschberg, D. S. 1982. Parallel graph algorithms without memory conflicts. In *Proceedings of the 20th Allerton Conference* (October), pp. 257–263. University of Illinois, Urbana-Champaign.

Hirschberg, D. S., Chandra, A. K., and Sarwate, D. V. 1979. Computing connected components on parallel computers. *Communications of the ACM* 22, 8 (August), pp. 461–464.

Hoare, C. A. R. 1962. Quicksort. *Computer Journal* 5, pp. 10–15.

Hoare, C. A. R. 1974. Monitors: An operating system structuring concept. *Communications of the ACM* 17, 10 (October), pp. 549–557.

Hoare, C. A. R. 1978. Communicating sequential processes. *Communications of the ACM* 21, 8 (August), pp. 666–677.

Hockney, R. W., and Jesshope, C. R. 1981. *Parallel Computers: Architecture, Programming and Algorithms.* Adam Hilger, Bristol, England.

Hoey, D., and Leiserson, C. E. 1980. A layout for the shuffle-exchange network. In *Proceedings of the 1980 International Conference on Parallel Processing* (August), pp. 329–336. IEEE, New York.

Holt, R. C., Graham, G. S., Lazowska, E. D., and Scott, M. A. 1978. *Structured Concurrent Programming with Operating System Applications.* Addison-Wesley, Reading, MA.

Hong, Z., and Sedgewick, R. 1982. Notes on merging networks. *Proceedings of the 14th Annual ACM Symposium on the Theory of Computing* (May), pp. 296–302. ACM, New York.

Hopcroft, J. E., and Ullman, J. D. 1973. Set-merging algorithms. *SIAM Journal on Computing* 2, pp. 294–303.

Hopcroft, J. E., and Ullman, J. D. 1979. *Introduction to Automata Theory, Languages, and Computation.* Addison-Wesley, Reading, MA.

Horowitz, E. 1979. VLSI architectures for matrix computations. In *Proceedings of the 1979 International Conference on Parallel Processing* (August), pp. 124–127. IEEE, New York.

Horowitz, E., and Sahni, S. 1978. *Fundamentals of Computer Algorithms.* Computer Science Press, Potomac, MD.

Horowitz, E., and Zorat, A. 1983. Divide and conquer for parallel processing. *IEEE Transac-*

tions on Computers C-32, 6 (June), pp. 582–585.

Hsiao, C. C., and Snyder, L. 1983. Omni-sort: A versatile data processing operation for VLSI. In *Proceedings of the 1983 International Conference on Parallel Processing* (August), pp. 222–225. IEEE, New York.

Hudak, P. 1986. Para-functional programming. *Computer* 19, 8 (August), pp. 60–70.

Hull, M. E. C. 1984. A parallel view of stable marriages. *Information Processing Letters* 18 (28 February), pp. 63–66.

Hwang, K. 1984. *Supercomputers: Design and Applications.* IEEE Computer Society Press, Silver Spring, MD.

Hwang, K., and Briggs, F. A. 1984. *Computer Architecture and Parallel Processing.* McGraw-Hill, New York.

Hwang, K., and Cheng, Y.-H. 1982. Partitioned matrix algorithms for VLSI arthmetic system. *IEEE Transactions on Computers* C-31, 12 (December), pp. 1215–1224.

Hwang, K., Su, S. P., and Ni, L. M. 1981. Vector computer architecture and processing techniques. In *Advances in Computers,* vol. 20, M. Yovits, ed. Academic Press, New York, pp. 115–197.

Hyafil, L. 1976. Bounds for selection. *SIAM Journal on Computing* 5, 1 (February), pp. 109–114.

Ibaraki, T. 1976a. Computational efficiency of approximate branch-and-bound algorithms. *Mathematical Operations Research* 1, 3, pp. 287–298.

Ibaraki, T. 1976b. Theoretical comparisons of search strategies in branch-and-bound algorithms. *International Journal of Computer and Information Science* 5, 4, pp. 315–344.

Imai, M., Fukumara, T., and Yoshida, Y. 1979. A parallelized branch-and-bound algorithm: Implementation and efficiency. *System Computer Controls* 10, 3, pp. 62–70.

Irani, K. B., and Chen, K.-W. 1982. Minimization of interprocessor communication for parallel computation. *IEEE Transactions on Computers* C-31, 11 (November), pp. 1067–1075.

Jacobi, C. G. J. 1845. Über eine Neue Auflösungsart der bei der Methode der Kleinsten Quadrate Vorkommenden Lineären Gleichungen. *Astr. Nachr.* 22 (523), pp. 297–306.

Ja'Ja', J., and Simon, J. 1982. Parallel algorithms in graph theory: Planarity testing. *SIAM Journal on Computing* 11, 2 (May), pp. 314–328.

Jess, J. A. G., and Kees, H. G. M. 1982. A data structure for parallel L/U decomposition. *IEEE Transactions on Computers* C-31, 3 (March), pp. 231–239.

Jesshope, C. R. 1980. Implementation of fast RADIX 2 transforms on array processors. *IEEE Transactions on Computers* C-29, 1 (January), pp. 20–27.

Johnson, D. S. 1983. The NP-completeness column: An ongoing guide. *Journal of Algorithms* 4, pp. 189–203.

Jones, A. K., and Gehringer, E. F., eds. 1980. The Cm* multiprocessor project: A research review. Tech. Rept. CMU-CS-80-131, Dept. of Computer Science, Carnegie-Mellon University, Pittsburgh.

Jones, A. K., and Schwarz, P. 1980. Experience using multiprocessor systems—A status report. *ACM Computing Surveys* 12, 2 (June), pp. 121–165.

Jones, N. D. 1975. Space-bounded reducibility among combinatorial problems. *Journal of Computer Science and System Sciences* 11, pp. 68–85.

Joseph, M., Prasad, V. R., end Natarajan, N. 1984. *A Multiprocessor Operating System.* Prentice-Hall, Englewood Cliffs, NJ.

Kasahara, H., and Narita, S. 1984. Practical multiprocessor scheduling algorithms for efficient parallel processing. *IEEE Transactions on Computers* C-33, 11 (November), pp. 1023–1029.

Kedem, Z. M., and Zorat, A. 1981. On relations between input and communication/comparison in VLSI (preliminary report). In *Proceedings of the 22d Annual Symposium on Foundations of Computer Science* (October), IEEE, New York, pp. 37–44.

Keller, R. M. 1976. Formal verification of parallel programs. *Communications of the ACM* 19, 7 (July), pp. 371–384.

Kleitman, D., Leighton, F. T., Lepley, M., and Miller, G. L. 1981. New layouts for the shuffle-exchange graph. In *Proceedings of the 13th Annual ACM Symposium on Theory of Computing* (May), pp. 278–292. ACM, New York.

Knuth, D. E. 1973. *The Art of Computer Programming*, vol. 3. Addison-Wesley, Reading, MA.

Knuth, D. E. 1976. Big omicron and big omega and big theta. *SIGACT News* (April–June), pp. 18–23.

Knuth, D. E., and Moore, R. W. 1975. An analysis of alpha-beta pruning. *Artificial Intelligence* 6, pp. 293–326.

Kogge, P. M., and Stone, H. S. 1973. A parallel algorithm for the efficient solution of a general class of recurrence equations. *IEEE Transactions on Computers* C-22, 8 (August), pp. 786–793.

Kosaraju, S. R. 1979. Fast parallel processing array algorithms for some graph problems. In *Proceedings of the 11th Annual ACM Symposium on Theory of Computing* (May), ACM, New York, pp. 231–236.

Kowalik, J. S., ed. 1985. *Parallel MIMD Computation: HEP Supercomputer and Its Applications*. M.I.T. Press, Cambridge, MA.

Kowalski, R. 1979. *Logic for Problem Solving*. North-Holland, New York.

Kozen, D. 1977. Complexity of finitely presented algebras. In *Proceedings of the 9th Annual ACM Symposium on Theory of Computing* (May), ACM, New York, pp. 164–177.

Kramer, M. R., and van Leeuwen, J. 1982. Systolic computation and VLSI. Tech. Rept. RUU-CS-82-9, Vakgroep Informatica, Rijksuniversiteit Utrecht.

Kruskal, C. P. 1982. Results in parallel searching, merging, and sorting. *Proceedings of the 1982 International Conference on Parallel Processing* (August), pp. 196–198. IEEE, New York.

Kruskal, C. P. 1983. Searching, merging and sorting in parallel computation. *IEEE Transactions on Computers* C-32, 10 (October), pp. 942–946.

Kruskal, J. B. 1956. On the shortest subtree of a graph and the traveling salesman problem. *Proceedings of the American Mathematical Society* 7 (February), pp. 48–50.

Kučera, L. 1982. Parallel computation and conflicts in memory access. *Information Processing Letters* 14, 2 (20 April), pp. 93–96.

Kuck, D. J. 1977. A survey of parallel machine organization and programming. *Computing Surveys* 9, 1 (March), pp. 29–59.

Kuck, D. J. 1978. *The Structure of Computers and Computations*, vol. 1, Wiley, New York.

Kuhn, R. H., and Padua, D. A. 1981. *Tutorial on Parallel Processing*. IEEE Computer Society Press, Los Angeles.

Kulkarni, A. V., and Yen, D. W. L. 1982. Systolic processing and an implementation for signal and image processing. *IEEE Transactions on Computers* C-31, 10 (October), pp. 1000–1009.

Kumar, M., and Hirschberg, D. S. 1983. An efficient implementation of Batcher's odd-even merge algorithm and its application in parallel sorting schemes. *IEEE Transactions on Computers* C-32, 3 (March), pp. 254–264.

Kumar, S. P., and Kowalik, J. S. 1984. Parallel factorization of a positive definite matrix on an MIMD computer. In *Proceedings of the 1984 International Conference on Parallel Processing* (August), pp. 417–424. IEEE, New York.

Kumar, V., and Kanal, L. 1983. A general branch-and-bound formulation for understanding and synthesizing AND/OR tree search procedures. *Artificial Intelligence* 21, pp. 179–198.

Kumar, V., and Kanal, L. 1984. Parallel branch-and-bound formulations for AND/OR tree search. *IEEE Transactions on Pattern Analysis and Machine Intelligence*, PAMI-6, 6 (November), pp. 768–778.

Kung, H. T. 1976. Synchronized and asynchronous parallel algorithms for multiprocessors. In *Algorithms and Complexity: New Directions and Recent Results*, J. F. Traub, ed. Academic Press, New York, pp. 153–200.

Kung, H. T. 1980. The structure of parallel algorithms. In *Advances in Computers*, vol. 19, M. Yovits, ed. Academic Press, New York, pp. 65–112.

Kung, H. T. 1982. Why systolic architectures? *Computer* 15, 1 (January), pp. 37–46.

Kung, H. T., and Lehman, P. L. 1980. Concurrent manipulation of binary search trees. *ACM Transactions on Database Systems* 5 (September), pp. 354–382.

Kung, H. T., and Leiserson, C. E. 1980. Systolic arrays for VLSI. In *Introduction to VLSI Systems*, C. Mead and L. Conway, eds. Addison-Wesley, Reading, MA, pp. 260–292.

Kung, S.-Y., Arun, K. S., Gal-Ezer, R. J., and Bhaskar Rao, D. V. 1982. Wavefront array processor: Languages, architecture, and applications. *IEEE Transactions on Computers* C-31, 11 (November), pp. 1054–1066.

Ladner, R. E. 1975. The circuit value problem is log space complete for *P*. *SIGACT News* 7, 1 (January), pp. 18–20.

Lai, T.-H., and Sahni, S. 1983. Anomalies in parallel branch-and-bound algorithms. In *Proceedings of the 1983 International Conference on Parallel Processing* (August), pp. 183–190. Also in *Communications of the ACM* 27, 6 (June 1984), pp. 594–602. IEEE, New York.

Lakshmivarahan, S., Dhall, S. K., and Miller, L. L. 1984. Parallel sorting algorithms. In *Advances in Computers*, vol. 23, M. C. Yovitts, ed. Academic Press, New York, pp. 295–354.

Lambiotte, J. J., Jr., and Korn, D. D. 1979. Computing the fast Fourier transform on a vector computer. *Math. Comput.* 33 (July), pp. 977–992.

Lamport, L. 1977. Proving the correctness of multiprocess programs. *IEEE Transactions on Software Engineering* SE-3, 7 (March), pp. 125–143.

Lang, H.-W., Schimmler, M., Schmeck, H., and Schröder, H. 1983. A fast sorting algorithm for VLSI. In *Proceedings of the 10th International Colloquium on Automata, Languages, and Programming*, pp. 408–419.

Lang, T., and Stone, H. S. 1976. A shuffle-exchange network with simplified control. *IEEE Transactions on Computing* C-25 (January), pp. 55–56.

Lawrie, D. H. 1975. Access and alignment of data in an array processor. *IEEE Transactions on Computers* C-24, 12 (December), pp. 1145–1155.

LeBlanc, T. J. 1986. Shared memory versus message-passing in a tightly-coupled multiprocessor: A case study. Butterfly Project Report 3, Computer Science Dept., University of Rochester, Rochester, NY.

Lee, D. T., Chang, H., and Wong, C. K. 1981. An on-chip compare/steer bubble sorter. *IEEE Transactions on Computers* C-30, 6 (June), pp. 396–405.

Lehman, P. L., and Yao, S. B. 1981. Efficient locking for concurrent operations on B-trees. *ACM Transactions on Database Systems* 6, 4 (December), pp. 650–670.

Leighton, F. T. 1981. New lower bound techniques for VLSI. In *Proceedings of the 22d Annual Symposium on Foundations of Computer Science* (October), IEEE, New York, pp. 1–12.

Leighton, F. T. 1983. *Complexity Issues in VLSI*. M.I.T. Press, Cambridge, MA.

Leighton, F. T. 1984. Tight bounds on the complexity of parallel sorting. In *Proceedings of the 16th Annual ACM Symposium on Theory of Computing* (May), pp. 71–80. ACM, New York.

Leiserson, C. E. 1980. Area efficient graph layouts. In *Proceedings of the 21st Annual Symposium on Foundations of Computer Science* (October), pp. 270–281. IEEE, New York.

Leiserson, C. E. 1983. *Area-Efficient VLSI Computation*. M.I.T. Press, Cambridge, MA.

Leiserson, C. E., and Saxe, J. B. 1981. Optimizing synchronous systems. In *Proceedings of the 22d Annual Symposium on Foundations of Computer Science* (October), IEEE, New York, pp. 23–36.

Levialdi, S. 1972. On shrinking binary picture patterns. *Communications of the ACM* 15, 1 (January), pp. 2–10.

Levialdi, S. 1985. A pyramid project using integrated technology. In *Integrated Technology for Parallel Image Processing*, S. Levialdi, ed. Academic Press, New York.

Levine, R. D. 1982. Supercomputers. *Scientific American* 246, 1 (January), pp. 118–135.

Levitt, K. N., and Kautz, W. T. 1972. Cellular arrays for the solution of graph problems. *Communications of the ACM* 15, 9 (September), pp. 789–801.

Li, G.-J., and Wah, B. W. 1984a. How to cope with anomalies in parallel approximate branch-and-bound algorithms. In *Proceedings of the National Conference on Artificial Intelligence*, pp. 212–215.

Li, G.-J., and Wah, B. W. 1984b. Computational efficiency of parallel approximate branch-and-bound algorithms. In *Proceedings of the 1984 International Conference on Parallel Processing* (August), pp. 473–480. IEEE, New York.

Li, G.-J., and Wah, B. W. 1985. MANIP-2: A multicomputer architecture for solving logic programming problems. In *Proceedings of the 1985 International Conference on Parallel Processing* (August), pp. 123–130. IEEE, New York.

Lincoln, N. R. 1982. Technology and design trade-offs in the creation of a modern supercomputer. *IEEE Transactions on Computers* C-31, 5 (May), pp. 349–362.

Lindstrom, G., and Panangaden, P. 1984. Stream-based execution of logic programs. In *Proceedings of the 1984 International Symposium on Logic Programming* (February), pp. 168–176.

Lint, B., and Agerwala, T. 1981. Communication issues in the design and analysis of parallel algorithms. *IEEE Transactions on Software Engineering* SE-7, 2 (March), pp. 174–188.

Lipton, R. J., and Valdes, J. 1981. Census functions: An approach to VLSI upper bounds (preliminary version). In *Proceedings of the 22d Annual Symposium on Foundations of Computer Science* (October), IEEE, New York, pp. 13–22.

Liskov, B. L., and Scheifler, R. 1982. Guardians and actions: Linguistic support for robust, distributed programs. In *Proceedings of the 9th ACM Symposium on Reliability in Distributed Software and Database Systems* (Pittsburgh, July 21–22). IEEE, New York, pp. 53–60.

Little, J. D. C., Murty, K. G., Sweeney, D. W., and Karel, C. 1963. An algorithm for the traveling salesman problem. *Operations Research* 11, 6 (November-December), pp. 972–989.

Lloyd, J. W. 1984. *Foundations of Logic Programming*. Springer-Verlag, New York.

Lord, R. E., Kowalik, J. S., and Kumar, S. P. 1983. Solving linear algebraic equations on an MIMD computer. *Journal of the ACM* 30, 1 (January), pp. 103–117.

Lorin, H. 1972. *Parallelism in Hardware and Software: Real and Apparent Concurrency*. Prentice-Hall, Englewood Cliffs, NJ.

Lorin, H. 1975. *Sorting and Sort Systems*. Addison-Wesley, Reading, MA.

Loui, M. C. 1984. The complexity of sorting on distributed systems. *Information and Control* 60, pp. 70–85.

Lubeck, O., Moore, J., and Mendez, R. 1984. A benchmark comparison of three computers: Fujitsu VP-200, Hitachi S810/20 and Cray X-MP/2. Preprint LA-UR-84-3584, Los Alamos National Laboratory, Los Alamos, NM.

Luk, F. T. 1980. Computing the singular-value decomposition on the ILLIAC IV. *ACM Transactions on Mathematical Software* 6, 4 (December), pp. 524–539.

Madnick, S. E., and Donovan, J. J. 1974. *Operating Systems*, McGraw-Hill, New York.

Manber, U., and Ladner, R. E. 1982. Concurrency control in a dynamic search structure. Technical Rept. 82-01-01, Dept. Computer Science, University of Washington, Seattle.

Marsland, T. A., and Campbell, M. 1982. Parallel search of strongly ordered game trees. *Computing Surveys* 14, 4 (December), pp. 533–551.

Marsland, T. A., and Rushton, P. G. 1974. A study of techniques for game-playing programs. In *Advances in Cybernetics and Systems*, vol. 1, J. Rose, ed. Gordon and Breach, London, pp. 363–371.

Martelli, A., and Montanari, U. 1973. Additive AND/OR graphs. *Proceedings of the International Joint Conference on Artificial Intelligence*, pp. 1–11.

Mashburn, H. H. 1979. The C.mmp/Hydra: An architectural overview. Tech. Rept., Dept. of Computer Science, Carnegie-Mellon University, Pittsburgh.

Mateti, P., and Deo, N. 1981. Parallel algorithms for the single source shortest path problem. Tech. Rept. CS-81-078, Computer Science Dept., Washington State University, Pullman.

Mead, C., and Conway, L. 1980. *Introduction to VLSI Systems*. Addison-Wesley, Reading, MA.

Mead, C., and Rem, M. 1979. Cost and performance of VLSI computing structures. *IEEE Journal on Solid State Circuits* SC-14, 2, pp. 455–462.

Meertens, L. G. L. T. 1979. Bitonic sort on ultracomputers. Tech. Rept. 117/79 (September), Dept. of Computer Science, The Mathematical Centre, Amsterdam.

Meggido, N. 1983. Applying parallel computation algorithms in the design of serial algorithms. *Journal of the ACM* 30, 4 (October), pp. 852–865.

Mierowsky, C., Taylor, S., Shapiro, E., Levy, J., and Safra, M. 1985. The design and implementation of flat Concurrent Prolog. Tech. Rept. CS85-09, Weizmann Inst. Science, Rehovot, Israel.

Miller, R., and Stout, Q. F. 1984a. Computational geometry on a mesh-connected computer. In *Proceedings of the 1984 International Conference on Parallel Processing* (August), pp. 66–73. IEEE, New York.

Miller, R., and Stout, Q. F. 1984b. Convexity algorithms for pyramid computers. In *Proceedings of the 1984 International Conference on Parallel Processing* (August), pp. 177–184. IEEE, New York.

Miller, R., and Stout, Q. F. 1985a. Geometric algorithms for digitized pictures on a mesh-connected computer. *IEEE Transactions on Pattern Analysis and Machine Intelligence*, PAMI-7, pp. 216–228.

Miller, R., and Stout, Q. F. 1985b. Pyramid computer algorithms for determining geometric properties of images. In *Proceedings of the 1985 ACM Symposium on Computational Geometry*, pp. 263–277.

Miller, R., and Stout, Q. F. 1986. Data movement techniques for the pyramid computer. *SIAM Journal on Computing* 15 (to appear).

Minsky, M., and Papert, S. 1971. On some associative parallel and analog computations. In *Associative Information Techniques*, E. J. Jacks, ed. American Elsevier, New York.

Miranker, G., Tang, L., and Wong, C. K. 1983. A "zero-time" VLSI sorter. *IBM Journal of Research and Development* 27, 2, pp. 140–148.

Misra, J., and Chandy, K. M. 1982. A distributed graph algorithm: Knot detection. *ACM Transactions on Programming Languages and Systems* 4, 4 (October), pp. 678–686.

Mohan, J. 1983. Experience with two parallel programs solving the traveling salesman problem. In *Proceedings of the 1983 International Conference on Parallel Processing* (August), pp. 191–193, IEEE, New York.

Moore, E. F. 1959. The shortest path through a maze. In *Proceedings of the International Symposium on the Theory of Switching* (held in 1957), vol. 2, pp. 285–292.

Moravec, H. P. 1979. Fully interconnected multiple computers with pipelined sorting nets. *IEEE Transactions on Computers* C-28, 10 (October), pp. 795–801.

Moto-oka, T., ed. 1982. *Fifth Generation Computer Systems*. North-Holland, New York.

Mukhopadhyay, A. 1981. WEAVESORT—A new sorting algorithm for VLSI. Tech. Rept. TR-53-81, University of Central Florida, Orlando.

Mukhopadhyay, A., and Ichikawa, T. 1972. An *n*-step parallel sorting machine. Tech. Rept. 72-03, Dept. of Computer Science, The University of Iowa.

Muller, D. E., and Preparata, F. P. 1975. Bounds to complexities of networks for sorting and for switching. *Journal of the ACM* 22, 2 (April), pp. 195–201.

Mundie, D. A., and Fisher, D. A. 1986. Parallel processing in Ada. *Computer* 19, 8 (August), pp. 20–25.

Nassimi, D., and Sahni, S. 1979. Bitonic sort on a mesh-connected parallel computer. *IEEE Transactions on Computers* C-28, 1 (January), pp. 2–7.

Nassimi, D., and Sahni, S. 1980a. An optimal routing algorithm for mesh-connected parallel computers. *Journal of the ACM* 27, 1 (January), pp. 6–29.

Nassimi, D., and Sahni, S. 1980b. Finding connected components and connected ones on a mesh-connected parallel computer. *SIAM Journal on Computing* 9, 4 (November), pp. 744–757.

Nassimi, D., and Sahni, S. 1981. Data broadcasting in SIMD computers. *IEEE Transactions on Computers* C-30, pp. 101–107.

Nassimi, D., and Sahni, S. 1982. Parallel permutation and sorting algorithms and a new generalized connection network. *Journal of the ACM* 29, 3 (July), pp. 642–667.

Nath, D., and Maheshwari, S. N. 1982. Parallel algorithms for the connected components and minimal spanning tree problems. *Information Processing Letters* 14, 1 (27 March), pp. 7–11.

Nath, D., Maheshwari, S. N., and Bhatt, P. C. P. 1983. Efficient VLSI networks for parallel processing based on orthogonal trees. *IEEE Transactions on Computers* C-32, 6 (June), pp. 569–581.

Nau, D. S. 1982. An investigation of the causes of pathology in games. *Artificial Intelligence* 19, pp. 257–278.

Oleinick, P. N. 1982. *Parallel Algorithms on a Multiprocessor*. UMI Research Press, Ann Arbor, MI.

Orcutt, S. E. 1974. Computer organization and algorithms for very high speed computations. Ph.D. dissertation, Stanford University, Stanford, CA.

Orenstein, J. A., Merrett, T. H., and Devroye, L. 1983. Linear sorting with $O(\log n)$ processors. *BIT* 23, pp. 170–180.

Ostlund, N. S., Hibbard, P. G., and Whiteside, R. A. 1982. A case study in the application of a tightly coupled multiprocessor to scientific computations. In *Parallel Computations*, G. Rodrigue, ed. Academic Press, New York, pp. 315–364.

Ottman, T. A., Rosenberg, A. L., and Stockmeyer, L. J. 1982. A dictionary machine (for VLSI). *IEEE Transactions on Computers* C-31, 9 (September), pp. 892–897.

Owicki, S., and Gries, D. 1976. Verifying properties of parallel programs: An axiomatic approach. *Communications of the ACM* 19, 5 (May), pp. 279–285.

Owicki, S., and Lamport, L. 1982. Proving liveness properties of concurrent programs. *ACM Transactions on Programming Languages and Systems* 4, 3 (July), pp. 455–495.

Paige, R. C., and Kruskal, C. P. 1985. Parallel algorithms for shortest path problems. In *Proceedings of the 1985 International Conference on Parallel Processing* (August), pp. 14–20. IEEE, New York.

Pape, U. 1974. Implementation and efficiency of Moore-algorithms for the shortest route problem. *Mathematical Programming* 7, 2 (October), pp. 212–222.

Parker, D. S., Jr. 1980. Notes on shuffle/exchange-type switching networks. *IEEE Transactions on Computers* C-29, 3 (March), pp. 213–222.

Paul, G. 1978. Large-scale vector/array processors. IBM Research Report RC 7306, September.

Pease, M. C. III. 1968. An adaption of the fast Fourier transform for parallel processing. *Journal of the ACM* 15, 2 (April), pp. 252–264.

Pease, M. C. III. 1977. The indirect binary n-cube microprocessor array. *IEEE Transactions on Computers* C-26, 5 (May), pp. 458–473.

Perl, Y. 1983. Bitonic and odd-even networks are more than merging. Tech. Rept., Rutgers University, New Brunswick, NJ.

Peters, F. 1981. Tree machine and divide-and-conquer algorithms. In *CONPAR '81, Lecture Notes in Computer Science* 111, Springer-Verlag, pp. 25–35.

Peterson, G. L. 1981. Myths about the mutual exclusion problem. *Information Processing Letters* 12, 3 (June), pp. 115–116.

Peterson, G. L. 1982. An $O(n \log n)$ unidirectional algorithm for the circular extrema problem. *ACM Transactions on Programming Language and Systems* 4, 4, pp. 758–762.

Potter, J. L. 1985. Programming the MPP. In *The Massively Parallel Processor*, J. L. Potter, ed. M.I.T. Press, Cambridge, MA, pp. 218–229.

Preparata, F. P. 1978. New parallel sorting schemes. *IEEE Transactions on Computers* C-27, 7 (July), pp. 669–673.

Preparata, F. P., and Vuillemin, J. 1981. The cube-connected cycles: A versatile network for parallel computation. *Communications of the ACM* 24, 5 (May), pp. 300–309.

Price, C. C. 1982. A VLSI algorithm for shortest path through a directed acyclic graph. *Congressus Numerantium* 34, pp. 363–371.

Price, C. C. 1983. Task assignment using a VLSI shortest path algorithm. Tech. Rept., Dept. of Computer Science, Stephen F. Austin State University, Nacogdoches, TX.

Prim, R. C. 1957. Shortest connection networks and some generalizations. *Bell System Technical Journal* 36, pp. 1389–1401.

Quinn, M. J. 1983. The design and analysis of algorithms and data structures for the efficient solution of graph theoretic problems on MIMD computers. Ph.D. dissertation, Computer Science Dept., Washington State University, Pullman.

Quinn, M. J. 1985. A note on two parallel algorithms to solve the stable marriage problem. *BIT* 25, pp. 473–476.

Quinn, M. J. 1987. Parallel sorting algorithms for tightly coupled multiprocessors. *Parallel Computing* (to appear).

Quinn, M. J., and Deo, N. 1983a. An approximate algorithm for the Euclidean traveling salesman problem. Tech. Rept. CS-83-105, Computer Science Dept., Washington State University, Pullman.

Quinn, M. J., and Deo, N. 1983b. An upper bound for the speedup of parallel branch-and-bound algorithms. In *Proceedings of the Third Conference on Foundations of Software Technology and Theoretical Computer Science* (December 1983), Bangalore, India, pp. 488–504.

Quinn, M. J., and Deo, N. 1984. Parallel graph algorithms. *Computing Surveys* 16, 3 (September), pp. 319–348.

Quinn, M. J., and Deo, N. 1986. An upper bound for the speedup of parallel best-bound branch-and-bound algorithms. *BIT* 26, 1 (March), pp. 35–43.

Quinn, M. J., and Yoo, Y. B. 1984. Data structures for the efficient solution of graph theoretic problems on tightly-coupled MIMD computers. In *Proceedings of the 1984 International Conference on Parallel Processing* (August), IEEE, New York, pp. 431–438.

Raghavendra, C. S., and Prasanna Kumar, V. K. 1986. *IEEE Transactions on Computers* C-35, 7 (July), pp. 662–669.

Ramakrishnan, I. V., and Varman, P. J. 1984. Modular matrix multiplication on a linear array. *IEEE Transactions on Computers* C-33, 11 (November), pp. 952–958.

Ramamoorthy, C. V., and Chang, L.-C. 1971. System segmentation for the parallel diagnosis of computers. *IEEE Transactions on Computers* C-20, 2 (February), pp. 153–161.

Ramamoorthy, C. V., Turner, J. L., and Wah, B. W. 1978. A design of a fast cellular associative memory for ordered retrieval. *IEEE Transactions on Computers* C-27, 9 (September), pp. 800–815.

Raskin, L. 1978. Performance evaluation of multiple processor systems. Ph.D. dissertation (August), Carnegie-Mellon University, Pittsburgh.

Reddaway, S. F. 1979. The DAP approach. In *Infotech State of the Art Report: Supercomputers*, vol. 2, C. R. Jesshope and R. W. Hockney, eds. Infotech, Maidenhead, England, pp. 311–329.

Redinbo, G. R. 1979. Finite field arithmetic on an array processor. *IEEE Transactions on Computers* C-28, 7 (July), pp. 461–471.

Reghbati (Arjomandi), E., and Corneil, D. G. 1978. Parallel computations in graph theory. *SIAM Journal on Computing* 2, 2 (May), pp. 230–237.

Reif, J. H. 1982. Symmetric complementation. In *Proceedings of the 14th Annual ACM Symposium on Theory of Computing* (May), pp. 201–214. ACM, New York.

Reif, J. H., and Spirakis, P. 1982. The expected time complexity of parallel graph and digraph algorithms. Tech. Rept. TR-11-82, Aiken Computation Laboratory, Harvard University, Cambridge, MA.

Reif, J. H., and Valiant, L. G. 1983. A logarithmic time sort for linear size networks. In *Proceedings of the 15th Annual ACM Symposium on Theory of Computing* (May), pp. 10–16. ACM, New York.

Reingold, E. M., Nievergelt, J., and Deo, N. 1977. *Combinatorial Algorithms: Theory and*

Practice. Prentice-Hall, Englewood Cliffs, NJ.

Reischuk, R. 1981. A fast probabilistic parallel sorting algorithm. In *Proceedings of the 22d Annual IEEE Symposium on Foundations of Computer Science* (October), pp. 212–219. IEEE, New York.

Ritchie, D. M., and Thompson, D. 1974. The UNIX timesharing system. *Communications of the ACM* 17, 7 (July), pp. 365–375.

Robinson, J. T. 1977. Analysis of asynchronous multiprocessor algorithms with applications to sorting. In *Proceedings of the 1977 International Conference on Parallel Processing* (August), pp. 128–135. IEEE, New York.

Robinson, J. T. 1979. Some analysis techniques for asynchronous multiprocessor algorithms. *IEEE Transactions on Software Engineering* (January), pp. 24–30.

Rodeheffer, T. L., and Hibbard, P. G. 1980. Automatic exploitation of parallelism on a homogeneous asynchronous multiprocessor. In *Proceedings of the 1980 International Conference on Parallel Processing* (August), pp. 15–16. IEEE, New York.

Rosenfeld, A. 1985. The prism machine: An alternative to the pyramid. *Journal of Parallel and Distributed Computing* 2, 4 (November), pp. 404–411.

Rosenkrantz, D., Stearns, R., and Lewis, P. 1974. Approximate algorithms for the traveling salesperson problem. In *Proceedings of the 15th Annual Symposium on Switching and Automata Theory* (October), pp. 33–42. IEEE, New York.

Rotem, D., Santoro, N., and Sidney, J. B. 1983. Distributed sorting. Tech. Rept. SCS-TR-#34 (December), School of Computer Science, Carleton University, Ottowa, Ontario, Canada.

Roussel, P. 1975. PROLOG: Manuel de reference et d'utilisation. Groupe Intelligence Artificielle, Universite d'Aix-Marseille, Luminy, September.

Rudolph, L. 1984. A robust sorting network. In *Proceedings of the 1984 Conference on Advanced Research in VLSI*, Massachusetts Institute of Technology (January), pp. 26–33.

Ruzzo, W., and Snyder, L. 1981. Minimum edge length planar embeddings of trees. In *VLSI Systems and Computations*, H. T. Kung, B. Sproull, and G. Steele, eds. Springer-Verlag, New York, pp. 119–123.

Satyanarayanan, M. 1980. Multiprocessing: An annotated bibliography. *IEEE Computer* 13, 5 (May), pp. 101–116.

Savage, C. 1977. Parallel algorithms for graph theoretic problems. Ph.D. dissertation, University of Illinois, Urbana.

Savage, C. 1981. A systolic data structure chip for connectivity problems. In *VLSI Systems and Computations*, H. T. Kung, R. F. Sproull, and G. L. Steele, Jr., eds. Computer Science Press, Rockville, MD.

Savage, C., and Ja'Ja', J. 1981. Fast, efficient parallel algorithms for some graph problems. *SIAM Journal on Computing* 10, 4 (November), pp. 682–690.

Savage, J. E. 1981. Planar circuit complexity and the performance of VLSI algorithms. In *VLSI Systems and Computations*, H. T. Kung, R. F. Sproull, and G. L. Steele, Jr., eds. Computer Science Press, Rockville, MD, pp. 61–66.

Schaefer, D. H., and Fisher, J. R. 1982. Beyond the supercomputer. *IEEE Spectrum* 19, 3 (March), pp. 32–37.

Schröder, H. 1983. Partition sorts for VLSI. *Informatik Fachberichte* 73, pp. 101–116.

Schwartz, J. T. 1980. Ultracomputers. *ACM Transactions on Programming Languages and Systems* 2, 4, pp. 484–521.

Sedgewick, R. 1983. *Algorithms.* Addison-Wesley, Reading, MA.

Seitz, C. L. 1985. The cosmic cube. *Communications of the ACM* 28, 1 (January), pp. 22–33.

Shapiro, E. Y. 1983a. The fifth generation project—A trip report. *Communications of the ACM* 26, 9 (September), pp. 637–641.

Shapiro, E. Y. 1983b. *Algorithmic Program Debugging.* M.I.T. Press, Cambridge, MA.

Shapiro, E. Y. 1983c. The Bagel: A systolic Concurrent Prolog machine (lecture notes). Weizmann Inst. Science, Rehovot, Israel.

Shapiro, E. Y. 1985. Systolic programming: A paradigm of parallel processing. Tech. Rept.

CS84-16, Weizmann Inst. Science, Rehovot, Israel.

Shapiro, E. Y. 1986. Concurrent Prolog: A progress report. *Computer* 19, 8 (August), pp. 44–58.

Shaw, A. C. 1974. *The Logical Design of Operating Systems.* Prentice-Hall, Englewood Cliffs, NJ.

Shiloach, Y., and Vishkin, U. 1981. Finding the maximum, merging and sorting in a parallel computation model. *Journal of Algorithms* 2, 1 (March), pp. 88–102.

Shiloach, Y., and Vishkin, U. 1982a. An $O(\log n)$ parallel connectivity algorithm. *Journal of Algorithms* 3, 1 (March), pp. 57–67.

Shiloach, Y., and Vishkin, U. 1982b. An $O(n^2 \log n)$ parallel MAX-FLOW algorithm. *Journal of Algorithms* 3, 2 (June), pp. 128–146.

Shröder, H. 1983. Partition sorts for VLSI. *Informatik Fachberichte* 73, pp. 101–116.

Siegel, H. J. 1977. The universality of various types of SIMD machine interconnection networks. In *Proceedings of the 4th Annual Symposium on Computer Architecture* (March), pp. 70–79. IEEE, New York.

Siegel, H. J. 1979a. Interconnection networks for SIMD machines. *Computer* 12, 6 (June), pp. 57–65.

Siegel, H. J. 1979b. A model of SIMD machines and a comparison of various interconnection networks. *IEEE Transactions on Computers* C-28, 12 (December), pp. 907–917.

Siegel, H. J. 1984. *Interconnection Networks for Large-Scale Parallel Processing: Theory and Case Studies.* Lexington Books, Lexington, MA.

Slagle, J. R., and Dixon, J. K. 1969. Experiments with some programs that search game trees. *Journal of the ACM* 16, 2 (April), pp. 189–207.

Smith, B. J. 1978. A pipelined shared resource MIMD computer. In *Proceedings of the 1978 International Conference on Parallel Processing* (August), pp. 6–8. IEEE, New York.

Snir, M. 1982. On parallel search. Presented at the Ottowa Conference on Distributed Computing (August).

Sollin, M. 1977. An algorithm attributed to Sollin. In *Introduction to the Design and Analysis of Algorithms*, S. E. Goodman and S. T. Hedetniemi, eds. McGraw-Hill, New York, sec. 5.5.

Stewart, G. W. 1973. *Introduction to Matrix Computations.* Academic Press, New York.

Stockman, G. 1979. A minimax algorithm better than alpha-beta? *Artificial Intelligence* 12, pp. 179–196.

Stone, H. S. 1971. Parallel processing with the perfect shuffle. *IEEE Transactions on Computers* C-20, 2 (February), pp. 153–161.

Stone, H. S. 1973. Problems of parallel computation. In *Complexity of Sequential and Parallel Numerical Algorithms*, J. F. Traub, ed. Academic Press, New York, pp. 1–16.

Stone, H. S. 1978. Sorting with STAR. *IEEE Transactions on Software Engineering* SE-4, 2 (February), pp. 138–146.

Stone, H. S. 1980. Parallel computers. In *Introduction to Computer Architecture*, H.S. Stone, ed. Science Research Associates, Chicago, chap. 8.

Stout, Q. F. 1983a. Sorting, merging, selecting and filtering on tree and pyramid machines. In *Proceedings of the 1983 International Conference on Parallel Processing* (August), pp. 214–221. IEEE, New York.

Stout, Q. F. 1983b. Mesh-connected computers with broadcasting. *IEEE Transactions on Computers* C-32, 9 (September), pp. 826–830.

Stout, Q. F. 1985a. Pyramid computer solutions of the closest pair problem. *Journal of Algorithms* 6, pp. 200–212.

Stout, Q. F. 1985b. Tree-based graph algorithms for some parallel computers (preliminary version). In *Proceedings of the 1985 International Conference on Parallel Processing* (August), pp. 727–730. IEEE, New York.

Swan, R. J., Bechtolsheim, A., Lai, K.-W., and Ousterhout, J. K. 1977. The implementation of the Cm* multi-microprocessor. In *Proceedings of the National Computer Conference*, AFIPS Press, Reston, VA, pp. 645–655.

Tanaka, Y., Nozaka, Y., and Masuyama, A. 1980. Pipeline searching and sorting modules as components of a data flow database computer. In *Proceedings IFIP Congress: Information Processing* 80, pp. 427–432.

Tanimoto, S. L. 1981. Towards hierarchical cellular logic: Design considerations for pyramid machines. Tech. Rept. 81-02-01, Department of Computer Science, University of Washington, Seattle.

Tanimoto, S. L. 1982a. Sorting, histogramming, and other statistical operations on a pyramid machine. Tech. Rept. 82-08-02, Dept. of Computer Science, University of Washington, Seattle.

Tanimoto, S. L. 1982b. Programming techniques for hierarchical parallel image processors. In *Multicomputers and Image Processing Algorithms and Programs*, K. Preston and L. Uhr, eds. Academic Press, New York, pp. 421–429.

Tanimoto, S. L., and Klinger, A. 1980. *Structured Computer Vision: Machine Perception through Hierarchical Computation Structures*, Academic Press, New York.

Thompson, C. D. 1979. Area-time complexity of VLSI. In *Proceedings of the 11th Annual ACM Symposium on Theory of Computing* (May), pp. 81–88. ACM, New York.

Thompson, C. D. 1980. A complexity theory for VLSI. Ph.D. dissertation, Dept. of Computer Science, Carnegie-Mellon University, Pittsburgh.

Thompson, C. D. 1983a. Fourier transforms in VLSI. *IEEE Transactions on Computers* C-32, 11 (November), pp. 1047–1057.

Thompson, C. D. 1983b. The VLSI complexity of sorting. *IEEE Transactions on Computers* C-32, 12 (December), pp. 1171–1184.

Thompson, C. D., and Kung, H. T. 1977. Sorting on a mesh-connected parallel computer. *Communications of the ACM* 20, 4 (April), pp. 263–271.

Thurber, K. J. 1976. *Large Scale Computer Architecture—Parallel and Associative Processors*. Hayden Book Co., Hasbrouck Heights, NJ.

Thurber, K. J. 1979a. Parallel processor architectures—Part I: General purpose systems. *Computer Design* (January), pp. 89–97.

Thurber, K. J. 1979b. Parallel processor architectures—Part II: Special purpose systems. *Computer Design* (February), pp. 103–114.

Tiwari, P. 1986. An efficient parallel algorithm for shifting the root of a depth first spanning tree. *Journal of Algorithms* 7, pp. 105–119.

Todd, S. 1978. Algorithms and hardware for a merge sort using multiple processors. *IBM Journal of Research and Development* 22, 5, pp. 509–517.

Tolub, S., and Wallach, Y. 1978. Sorting on an MIMD-type parallel processing system. *Euromicro Journal* 4, pp. 155–161.

Tseng, S. S., and Lee, R. C. T. 1984a. A new parallel sorting algorithm based upon min-mid-max operations. *BIT* 24, pp. 187–195.

Tseng, S. S., and Lee, R. C. T. 1984b. A parallel algorithm to solve the stable marriage problem. *BIT* 24, pp. 308–316.

Tsin, Y. H., and Chin, F. Y. 1982. Efficient parallel algorithms for a class of graph theoretic problems. Tech. Rept., Dept. of Computing Science, University of Alberta, Edmonton, Alberta.

Uhr, L. 1972. Layered "recognition cone" networks that preprocess, classify, and describe. *IEEE Transactions on Computers* C-21, 7 (July), pp. 758–768.

Uhr, L. 1984. *Algorithm-Structured Computer Arrays and Networks*. Academic Press, Orlando, FL.

Ullman, J. D. 1984. *Computational Aspects of VLSI*. Computer Science Press, Rockville, MD.

USA Today. 1985. Computer aids wind shear study. *USA Today* (September 23, 1985), p. 8B.

U.S. Department of Defense. 1981. *Programming language Ada: Reference Manual*, vol. 106, *Lecture Notes in Computer Science*. Springer-Verlag, New York.

Valiant, L. G. 1975. Parallelism in comparison problems. *SIAM Journal on Computing* 3, 4 (September).

Valiant, L. G. 1981. Universality considerations in VLSI circuits. *IEEE Transactions on Computers* C-30, 2 (February), pp. 135–140.

van Scoy, F. L. 1976. Parallel algorithms in cellular spaces. Ph.D. dissertation, School of Engineering and Applied Science, University of Virginia, Charlottesville.

van Scoy, F. L. 1980. The parallel recognition of classes of graphs. *IEEE Transactions on Computers* C-29, 7 (July), pp. 563–570.

van Voorhis, D. C. 1971. On sorting networks. Ph.D. dissertation, Stanford University, Stanford, CA.

van Wijngaarden, A., Mailloux, B. J., Peck, J. L., Koster, C. H. A., Sintzoff, M., Lindsey, C. H., Meertens, L. G. L. T., and Fisker, R. G. 1975. Revised report on the algorithm language ALGOL68. *Acta Informatica* 5, 1–3, pp. 1–236.

Vishkin, U. 1983. Implementation of simultaneous memory access in models that forbid it. *Journal of Algorithms* 4, 1 (March), pp. 45–50.

Vishkin, U., and Wigderson, A. To appear. Trade-offs between width and depth in parallel computation. *SIAM Journal on Computing.*

Vuillemin, J. E. 1983. A combinatorial limit to the computing power of VLSI circuits. *IEEE Transactions on Computers* C-32, 3 (March), pp. 294–300.

Wah, B. W., and Chen, K. L. 1984. A partitioning approach to the design of selection networks. *IEEE Transactions on Computers* C-33, 3 (March), pp. 261–268.

Wah, B. W., Li, G.-J., and Yu, C. F. 1984. The status of MANIP—A multicomputer architecture for solving combinatorial extremum-search problems. In *Proceedings of the 11th Annual International Symposium on Computer Architecture* (March), pp. 56–63. IEEE, New York.

Wah, B. W., Li, G., and Yu, C. F. 1985. Multiprocessing of combinatorial search problems. *Computer* 18, 6 (June), pp. 93–108.

Wah, B. W., and Ma, E. Y. W. 1984. MANIP—A multicomputer architecture for solving combinatorial extremum search problems. *IEEE Transactions on Computers* C-33, 5 (May), pp. 377–390.

Wah, B. W., and Yu, C. F. 1982. Probabilistic modeling of branch-and-bound algorithms. *Proceedings of COMPSAC* (November), pp. 647–653.

Warren, D. H. D., Pereira, L. M., and Pereira, F. 1977. PROLOG—The language and its implementation compared with LISP. *SIGPLAN Notices* 12, 8. Also in *SIGART Newsletter* 64 (August 1977), pp. 109–115.

Warshall, S. 1962. A theorem on Boolean matrices. *Journal of the ACM* 9, 1 (January), pp. 11–12.

Watson, I., and Gurd, J. 1981. A practical data flow computer. *Computer* 14, 2 (February), pp. 51–57.

Wegner, L. M. 1982. Sorting a distributed file in a network. In *Proceedings of the 1982 Conference on Information Science Systems*, Princeton, NJ (March), pp. 505–509.

Weide, B. W. 1981. Analytical models to explain anomalous behavior of parallel algorithms. In *Proceedings of the 1981 International Conference on Parallel Processing* (August), IEEE, New York, pp. 183–187.

Winslow, L. E., and Chow, Y.-C. 1981. Parallel sorting machines: Their speed and efficiency. In *Proceedings of the AFIPS 1981 National Computer Conference*, pp. 163–165.

Winslow, L. E., and Chow, Y.-C. 1983. The analysis and design of some new sorting machines. *IEEE Transactions on Computers* C-32, 7 (July), pp. 677–683.

Wirth, N. 1976. *Algorithms + Data Structures = Programs.* Prentice-Hall, Englewood Cliffs, NJ.

Wirth, N. 1977a. Modula: A language for modular multiprogramming. *Software Practice and Experience* 7, pp. 33–35.

Wirth, N. 1977b. The use of Modula. *Software Practice and Experience* 7, pp. 37–65.

Wirth, N. 1977c. Design and implementation of Modula. *Software Practice and Experience* 7, pp. 67–84.

Wold, E. H., and Despain, A. M. 1984. Pipeline and parallel-pipeline FFT processors for VLSI implementations. *IEEE Transactions on Computers* C-33, 5 (May), pp. 414–426.

Wong, C. K., and Chang, S.-K. 1974. Parallel generation of binary search trees. *IEEE Transactions on Computers* C-23, 3 (March), pp. 268–271.

Wong, F. S., and Ito, M. R. 1984. Parallel sorting on a re-circulating systolic sorter. *Computer Journal* 27, 3, pp. 260–269.

Wu, C. L., and Feng, T. Y. 1981. Universality of the shuffle exchange network. *IEEE Transactions on Computers* C-30, 5 (May), pp. 324–331.

Wyllie, J. C. 1979. The complexity of parallel computations. Ph.D. dissertation, Dept. of Computer Science, Cornell University, Ithaca, NY.

Yasuura, H., Tagaki, N., and Yajima, S. 1982. The parallel enumeration sorting scheme for VLSI. *IEEE Transactions on Computers* C-31, 12 (December), pp. 1192–1201.

Yau, S. S., and Fung, H. S. 1977. Associative processor architecture—A survey. *Computing Surveys* 9, 1 (March), pp. 3–27.

Yoo, Y. B. 1983. Parallel processing for some network optimization problems. Ph.D. dissertation, Computer Science Dept., Washington State University, Pullman.

Yu, C. F., and Wah, B. W. 1983. Virtual-memory support for branch-and-bound algorithms. In *Proceedings Compsac* (November), pp. 618–626.

Yu, C. F., and Wah, B. W. 1984. Efficient branch-and-bound algorithms on a two-level memory hierarchy. In *Proceedings Compsac* (November), pp. 504–514.

Zhang, C. N., and Yun, D. Y. Y. 1984. Multi-dimensional systolic networks for discrete Fourier transform. In *Proceedings of the Eleventh International Symposium on Computer Architecture, SIGARCH Newsletter*, pp. 215–222.

INDEX